Primer on Printed Flexible Sensors

Primer on Printed Flexible Sensors

Júlio C. Viana

Primer on Printed Flexible Sensors

Revolutionizing Sensing Technologies

Springer

Júlio C. Viana
Department of Polymer Engineering
University of Minho
Guimarães, Portugal

ISBN 978-3-032-06553-7 ISBN 978-3-032-06554-4 (eBook)
https://doi.org/10.1007/978-3-032-06554-4

This Springer imprint is published by the registered company Springer Nature Switzerland AG
The registered company address is: Gewerbestrasse 11, 6330 Cham, Switzerland

Preface

Printed flexible sensors (PFS) represent an innovative class of sensing devices manufactured using printing techniques on flexible substrates. PFS are an emerging class of sensor technology that uses flexible substrates and functional inks to enable lightweight, low-cost, and adaptable sensing solutions. These sensors play a crucial role in various industries, including automotive, healthcare, wearables, robotics, and environmental monitoring. As technology advances, PFS are poised to revolutionize smart sensing, enabling next-generation applications of flexible electronics across industries.

This book provides an overview of PFS, introducing the fundamentals of PFS in terms of materials, printing technologies, and design considerations. A short introduction on types of sensors is given initially. Then, the main types of PFS are presented (strain, force and pressure, proximity, temperature, humidity, optical, gas and chemical, biosensors, OFET sensors, SAW sensors), comparing main sensing mechanisms and identifying main design key parameters. The book ends by discussing main technological challenges, and by identifying emerging technologies and trends and future prospects.

The book is organized into five chapters:

Introduction to Printed Flexible Sensors

Printed flexible sensors (PFS) represent an innovative class of sensing devices manufactured using printing techniques of functional inks on flexible substrates. This section categorizes the main types of sensors, their significance, and key applications. It also introduces the concept of figures of merit, which evaluate and compare the performance of PFS. A review of existing literature, mainly published review articles, highlights the progress and research focus in this field.

Fundamentals of Printed Flexible Sensors

The fundamentals of PFS involve selecting appropriate substrates for flexibility, using advanced functional inks, and employing various fabrication techniques such as screen printing, inkjet printing, and roll-to-roll manufacturing. Ensuring high-quality production requires rigorous testing and quality control, while the integration with electronic components involves specialized connectors, bonding techniques, and hybrid electronic systems. This section delves into the fundamental aspects of PFS, including the materials and methods used in their fabrication. It discusses:

- main substrates for flexibility;
- functional inks and materials (conductive, semiconductors, and dielectric inks);
- fabrication techniques (pre-printing, printing, and post-printing processes);
- testing and quality control methods to ensure reliability;
- advanced manufacturing techniques, including in-line and roll-to-roll production, and additive manufacturing.

Design Considerations

Design considerations for PFS focus on optimizing sensing mechanisms, form factors, flexibility, stretchability, and sustainability. Various sensing mechanisms, including piezoresistive, piezoelectric, capacitive, and electrochemical approaches, are used to detect different physical and chemical parameters. This section outlines the core design principles for PFS:

- sensing mechanisms (working principles of different sensors);
- design parameters (form factor, flexibility, stretchability, and integration strategies);
- interconnects design;
- sustainability challenges (eco-friendly materials and recyclability).

Types of Printed Flexible Sensors

There are multiple types of PFS, each one tailored to specific applications. A comprehensive classification of PFS is provided, covering multiple sensor types: strain sensors (piezoresistive, piezoelectric, capacitive), force and pressure sensors, proximity sensors (capacitive, resistive, inductive, optical, magnetic), temperature sensors (thermocouples, RTD, thermistors, capacitive, semiconductor-based, and optical sensors), humidity sensors (piezoresistive and capacitive mechanisms), optical sensors (types, materials, and key features), gas and chemical sensors (chemiresistive, capacitive, electrochemical, optical), biosensors (electrochemical, optical, piezoelectric), OFET—organic field-effect transistor sensors (design and material considerations), SAW—surface acoustic wave sensors (mechanisms, design, and materials). Each type of sensor is analyzed based on its sensing principles, key performance parameters, and comparative advantages.

Recent Advances, Challenges, and Future Trends

Despite their potential, PFS faces challenges related to material innovations, manufacturing scalability, and integration with existing technologies. Performance, reliability, and cost-effectiveness remain key concerns. However, advances in artificial intelligence (AI), the Internet of Things (IoT), and sustainable materials and manufacturing processes are driving new opportunities for PFS development. Looking ahead, PFS are expected to revolutionize sensor technology by enabling smart, flexible, and efficient solutions for a wide range of applications. As research continues, improvements in materials, manufacturing techniques, and sensor integration will further expand the capabilities and adoption of PFS. This section highlights the evolving landscape of PFS, focusing on:

- technological challenges (materials, manufacturing, integration, new sensor development);
- performance, reliability, and durability (strategies for improvement);
- scalability and cost-effectiveness (challenges and solutions);
- sustainability (eco-friendly practices and future trends);
- integration with AI and IoT (how PFS, AI, and IoT enhance smart applications);
- emerging Trends and innovations (breakthroughs driving the future of PFS).

Guimarães, Portugal
July 2025

Júlio C. Viana

[illegible] (OFET)—organic field-effect transistor sensors (design and material considerations); SAW—surface acoustic wave sensors (mechanisms, design, and materials). Each type of sensor is analyzed [illegible]

[illegible]

Acknowledgements

The content of this book has been shaped by research, reflection, and—perhaps most importantly—the people I've had the privilege to work with over the past decade. I am deeply grateful to all those who have shared their knowledge, challenged my thinking, or simply stood by me throughout this journey. This book would not have come to life without their support.

I would also like to thank Springer Nature for believing in this project and for providing the opportunity to bring it to publication.

Above all, I want to express my heartfelt thanks to my family. Their patience, understanding, and steady support—not only during the writing of this book, but through all the ups and downs of academic life—have meant more than I can say.

Contents

About the Author

Júlio C. Viana is an Associate Professor at the Department of Polymer Engineering at the University of Minho in Portugal. His research is carried out in IPC/LASI—Institute for Polymers and Composites/Associated Laboratory in Intelligent Systems, mainly related to advanced and multifunctional polymer systems, printed polymers, multifunctional and smart products, and non-conventional polymer technologies. Recently, he investigates large area/scattered electronics, structural electronics, flexible and stretchable electronics taking advantage of polymer properties, exploring the integration of smart materials in smart connected products. He is also currently President of DTx CoLab, an academia-industry collaborative laboratory in the digital transformation. Further information about his work and publications can be found here: https://www.cienciavitae.pt//pt/5F1B-D6BB-563F.

List of Figures

List of Tables

Chapter 1
Introduction to Printed Flexible Sensors

Printed flexible sensors, PFS, are revolutionizing the world of electronics by combining the advantages of flexible materials and printing technologies. PFS are devices created using printing techniques to deposit sensor materials (e.g., functional inks) on flexible substrates such as plastics, fabrics, or paper. PFS detect physical, chemical, or biological signals and convert them into electrical signals for data analysis. Unlike traditional rigid sensors, these sensors can bend, twist, and stretch, offering a unique solution in applications requiring adaptability to non-flat surfaces or dynamic environments. These sensors are lightweight, conformable, and can be produced cost-effectively, making them ideal for a wide range of applications, from wearable devices and healthcare monitoring to environmental sensing and smart packaging.

The field of PFS is rapidly evolving, driven by advancements in materials science, nanomaterials, and printing technologies [1, 2]. Future developments may lead to sensors with higher sensitivity, improved durability, and multifunctional capabilities. As technology advances, PFS are expected to become integral components on the Internet of Things (IoT), Industry 4.0, smart cities and mobility, and personalized healthcare, enhancing both our daily lives and industrial processes.

1.1 Main Types of Sensors

Sensors are devices that detect and measure physical or chemical quantities and convert them into electrical, optical, or mechanical signals that can be analyzed, processed, or displayed. There are numerous types of sensors, with various forms and configurations, depending on the specific application and sensing requirements. Some common types of sensors are shown in Table 1.1.

J. C. Viana, *Primer on Printed Flexible Sensors*,
https://doi.org/10.1007/978-3-032-06554-4_1

Table 1.1 Main types of sensors and applications

Sensor type	Physical measures	Main transduction principles	Main applications
Strain sensors	Measure strain or mechanical deformation	Resistive, capacitive, piezoelectric, surface acoustic waves (SAWs), optical	Structural health monitoring, wearable devices, and robotics
Force sensors	Measure the force applied to them	Piezoresistive, piezoelectric, capacitive, optical	Medical devices, automotive systems, touch interfaces
Pressure sensors	Detect pressure variations	Piezoresistive, piezoelectric, capacitive, resonant, optical	Medical devices, automotive systems, touch interfaces
Acoustic, Sound sensors	Detect and measure acoustic signals or sound waves	Capacitive, piezoelectric, light sensitive resistor (LDR), resistive, magnetic, SAW	Security systems, environment monitoring, home automation, automotive
Temperature sensors	Measure temperature changes in their surroundings	Thermocouples, resistance temperature detectors, thermistors, optical	Environmental monitoring, industrial processes, and consumer electronics
Humidity sensors	Measure humidity levels in the environment	Capacitive, resistive, gravimetric, optical	VAC systems, weather monitoring, agricultural
Gas, chemical sensors	Detect gases or chemicals	Electrochemical, resistance, photoionization, IR, SAW	Environmental monitoring, industrial safety, healthcare
Biosensors	Detect the presence of biological substances	Optical, electrochemical, piezoelectric, thermal	Medical diagnostics, food safety, environmental monitoring
Biometric	Measure physiological characteristics or behavioral patterns	Optical, pressure, capacitive, ultrasonic, acoustic, thermal, RF	Access control and security, mobile devices, vehicle access, border control, wearables
Biomechanical	Monitor body movements, muscle activity, etc	Capacitive, optical, ultrasonic, electrophysiological, motion and inertial force	Sports performance analysis, physical therapy, ergonomics
Light, Optical	Detect light or electromagnetic radiation	Photodiodes, photovoltaic cells, phototransistors, LDR	Equipment, environmental monitoring, biomedical imaging

(continued)

Table 1.1 (continued)

Sensor type	Physical measures	Main transduction principles	Main applications
Position	Measure the position or displacement of objects	Potentiometric, inductive, capacitive, optical, magnetic, piezoelectric	Labeling control, positioning system, robotics, appliances, automotive, aerospace
Motion	Detect movement or changes in position	Accelerometers, gyroscopes, optical, magnetometers	Vehicle dynamics, aerospace, robotics, equipment, navigation systems, stabilization platforms
Level	Measure the level of liquids or solids in tanks or containers	Float, capacitance, inductive, optical, ultrasonic, magneto-resistive	Industrial processes, medicine, healthcare, agriculture, waste material handling
Presence, proximity	Detect the presence or absence of objects within a certain range	Inductive, magnetic, ultrasonic, capacitive, IR/PIR, optical, radar, microwaves	Home automation, industrial automation, retail, security systems, automotive

A sensor device is composed of several elements (Fig. 1.1) that produce an output signal for the purpose of detecting a physical variable, its main components being typically:

- **Sensing Element**—directly interacts with the environment to detect the target parameter (detection); it undergoes physical or chemical changes in response to the measured parameter, generating an electrical, optical, or mechanical signal.
- **Transducer**—converts the output signal from the sensing element into a measurable electrical signal (conversion), which can be processed and analyzed by electronic circuits or systems (e.g., amplifiers, converters, signal conditioning circuits).

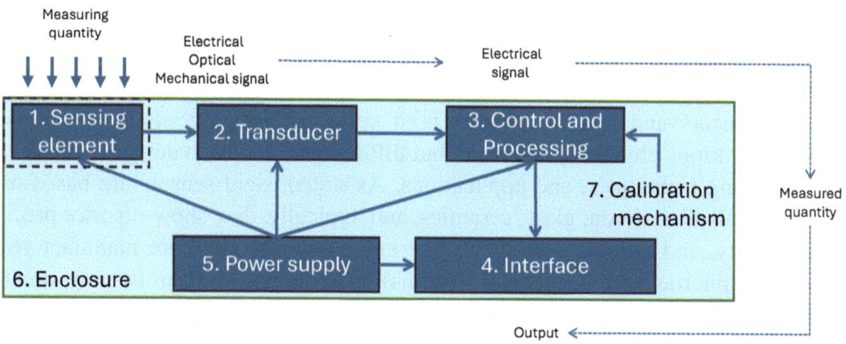

Fig. 1.1 Main components of sensor devices

- **Control and Processing**—to perform functions such as signal conditioning, filtering, linearization, averaging, calibration, and digital signal processing (signal processing); these circuits enhance the accuracy, stability, and performance of the sensor.
- **Interface**—where the measured data or signal is stored and made available to external systems or users (analog signals, digital signals, frequency outputs, communication interfaces (I2C, SPI, UART), or wireless protocols).
- **Power Supply**—to operate the sensor, providing electrical energy to power the sensing element, transducer, and associated electronics.
- **Enclosure**—provides physical protection to the sensing element and electronics, shielding them from environmental factors such as humidity, dust, temperature fluctuations, and mechanical shocks.
- **Calibration Mechanism**—needed to ensure accurate and reliable measurements over time (e.g., trimming potentiometers, calibration curves, software-based calibration algorithms).

1.2 Printed Flexible Sensors and Their Significance

PFS is a technology that combines flexible materials and printed electronics to develop flexible sensors (Fig. 1.2). PFS are manufactured using printing techniques onto flexible substrates, such as polymers, paper, or textiles. The printing processes used to manufacture these sensors typically involve techniques like screen printing, inkjet printing, flexographic printing, or roll-to-roll manufacturing. Main developments on printed electronics have been in printing sensing elements over flexible substrates (sensors). Nevertheless, printed electronic technologies are being used to fabricate other components of electronic devices (e.g., transistors, controllers, memories, batteries). Also, flexible hybrid electronics are being adopted to integrate both printed and advanced CMOS-based components on a flexible substrate, producing electronic devices that are flexible, stretchable, or conformable.

The requirements of PFS define the functional and non-functional characteristics that they should exhibit to fulfill their intended purpose effectively and efficiently. They shall meet the needs of a particular application or set of users, being application/product specific. Nevertheless, a set of main requirements of PFS can be identified, as shown in Table 1.2.

Rigid sensors and flexible sensors both serve the purpose of detecting and measuring various physical parameters but differ significantly in design, materials, manufacturing techniques, and applications. As stated, rigid sensors are based on rigid materials like silicon, glass, ceramics, and, typically, they show superior precision, stability, and durability in controlled environments. They are manufactured by MEMS (microelectromechanical systems) technologies and can be very small, millimeter to micron sized. Whereas, flexible sensors are made from flexible materials like polymers, textiles, paper, stretchable inks and can be very thin and have large coverage area. They are ideal for wearables, healthcare and applications requiring

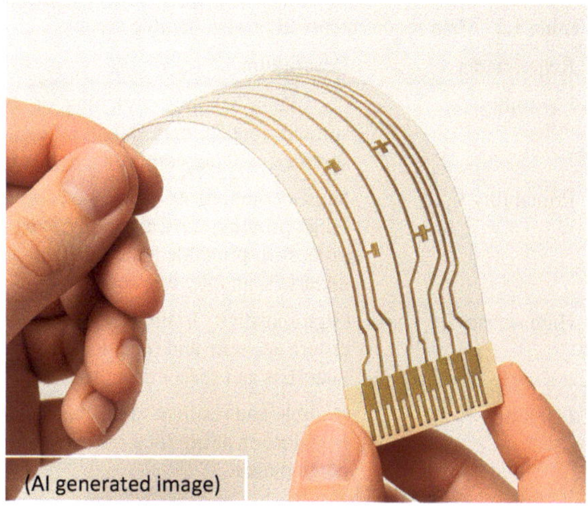

Fig. 1.2 Printed flexible sensors (PFS)

adaptability and lightweight designs. Table 1.3 compares both types of rigid and flexible sensors.

Rigid sensors are typically built from materials like silicon, glass, or ceramics. These materials provide high precision and stability, making rigid sensors ideal for environments requiring consistent and accurate measurements. They are often designed using MEMS, which provide miniaturized and reliable sensor solutions. Rigid sensors rely heavily on traditional semiconductor fabrication methods such as lithography, etching, and wafer bonding. These processes offer precision at the micro- and nanoscale, ensuring excellent accuracy but often require expensive manufacturing techniques with high precision class cleanroom environments. Rigid sensors are superior in precision, stability, and response time. Due to their solid structure, they are highly effective in environments with minimal mechanical stress.

In contrast, flexible sensors are made from flexible materials like polymers (e.g., PDMS, PET), graphene, or conductive inks. These sensors are lightweight and capable of bending, stretching, and conforming to various surfaces. Flexible sensor designs are often integrated with textiles, wearables, or flexible displays for applications that require adaptability and comfort. They are manufactured using simple techniques like screen printing, inkjet printing, and roll-to-roll manufacturing. These methods allow cost-effective, large-area sensor production, making flexible sensors suitable for mass-market applications like fitness trackers, electronic skin, and smart textiles. Flexible sensors, while offering greater adaptability, are slightly more susceptible to environmental damage, and they overperform in dynamic applications requiring movement, such as wearable health monitors or prosthetic devices. PFS offer several advantages over traditional rigid sensors, including

Table 1.2 Main requirements of printed flexible sensors

Requirement	Description
Flexibility	Capable of bending, stretching, or conforming seamlessly to irregular surfaces (e.g., human skin, textile fabric), to curved surfaces, or to soft surfaces, without compromising its functionality or performance
Printability	Sensor components should be compatible with printing processes (e.g., inkjet printing, screen printing, flexographic printing) and post-printing processes; printable materials include conductive inks, dielectric materials, and flexible substrates
High sensitivity	High sensitivity to the physical/chemical parameter being measured to ensure accurate and reliable detection; it often depends on the choice of materials and sensor design
Durability	Durable and resilient to mechanical loadings, such as bending, twisting, abrasion, ensuring long-term reliability and stability of sensor performance
Low cost	Manufactured using cost-effective printing techniques and materials to enable mass production at a low cost per unit
Environmental stability	Keep its performance characteristics (constancy) over a wide range of environmental conditions, e.g., temperature, humidity, exposure to chemicals
Environmental compatibility	Ability to interact with its surrounding environment without causing harm, adverse effects, or negative impact; chemical resistance (to chemicals, solvents, corrosive substances), temperature tolerance, humidity resistance, UV exposure, electromagnetic interference, mechanical durability (withstand shocks, vibrations)
Low-power consumption	Energy-efficient operation (e.g., battery-powered, portable devices) prolongs battery life and reduces the need for frequent battery replacements or recharging
Fast response time	Rapid detection and response to changes in the measured parameter are essential for real-time monitoring and control applications
Customizability	Easily customizable in terms of sensor geometry, size, and sensing characteristics to meet the specific requirements of different applications
Biocompatibility	Ability to interact with living systems without causing adverse effects; relevant if sensor is used in medical applications in contact with a living system
Compatibility with electronics	Compatible with the other sensor components; this includes integration with flexible or rigid electronic components

- **Flexibility and conformability**—printed sensors are thin, with ability to conform to various shapes and to irregular surfaces.
- **Lightweight**—due to the use of low-density materials and very thin substrates;
- **Customizable shape**—the flexible form factor enables custom designs to fit specific applications.
- **Large area sensor capabilities**—printed sensors can cover large surface areas, providing extensive sensing capabilities.

Table 1.3 Comparison between rigid and printed flexible sensors

Feature/Property	Rigid sensor	Flexible sensor
Substrates materials	Silicon, glass, ceramics, or other hard material	Polymers, paper, textiles, or thin-film materials; graphene, CNT, conductive inks
Young modulus	100–200 GPa	0.1–10 GPa
Stretchability	<1%	<500%
Conformable to non-flat surfaces	Poor	Excellent
Manufacturing method	Complex micro/nanofabrication processes; MEMS techniques	Printing techniques; easier to manufacture in large formats
Scale	0.01–0.1 m^2 (wafer)	1–100 m^2 (web)
Cleanroom class	High precision (class 1–10)	Lower requirement (class 100–1000)
Cost per unit area	High; often more expensive due to precise manufacturing processes	Low; generally cheaper when mass-produced due to simpler materials and production techniques
Energy requirement	Higher for precision fabrication	Lower due to simpler techniques, and low-temperature processes
Carbon footprint	High due to intensive processes	Lower with eco-friendly materials
Form factor	Generally bulkier and heavier	Thin, lightweight, and adaptable to various shapes
Accuracy and precision	Superior accuracy in static environments	Slightly lower precision but excel in dynamic conditions
Power efficiency	Often more power-efficient in stationary environments	Require energy optimization techniques due to stretching or movement
Integration	PCB-integrated Design; suitable for rigid surfaces and structured installations	Easily integrated into wearable devices, textiles, and curved surfaces

- **Scalability**—they can be produced using scalable manufacturing processes such as roll-to-roll printing, allowing for high-volume production with minimal labor and material costs, making them suitable for mass-market applications.
- **Cost-effective mass production**—reduced cost of materials, and cost-effective printing processes suitable for large-scale manufacturing.
- **Portability**—being lightweight and flexible, printed flexible sensors are highly portable; they can be easily rolled up or folded.
- **Environmental friendliness**—Printing techniques for flexible sensor fabrication often use environmentally friendly materials and processes (e.g., low temperature, no high-level cleanroom needed) compared to traditional sensor manufacturing

methods; additionally, the ability to print sensors directly onto substrates reduces material waste and energy consumption.
- **Integration**—PFS are compatible with flexible substrates, which allows for their seamless integration directly into the surface of the object they are sensing, eliminating the need for additional mounting hardware; this simplifies the design and reduces assembly complexity.
- **Fast Prototyping**—The ease of customization and low-cost fabrication make printed flexible sensors ideal for rapid prototyping and iterative design processes.
- **Application diversity**—Various types of PFS suitable for several applications.

These advantages and their versatility make PFS a promising technology for a wide range sensing of applications.

1.3 Main Applications

PFS are a rapidly growing area in sensor technology, offering numerous advantages such as lightweight, flexibility, scalability, and cost-effectiveness. They can be easily embedded in materials and products. These sensors find applications across various fields. They enable the development of innovative products and systems for a wide range of applications, including wearable technology, Internet of Things (IoT) devices, smart textiles, wearables, healthcare monitoring systems, automotive, aerospace, industrial monitoring, and consumer electronics. These applications (Table 1.4) demonstrate the versatility and utility of PFS across a wide range of industries and technological areas, enabling innovative products, processes, and services.

1.4 Figure of Merit of Printed Flexible Sensors

Figures of merit (FOM) are quantitative metrics used to evaluate and compare the performance of sensors objectively. They provide standardized means of assessing sensor capabilities (e.g., sensitivity, accuracy, response time), being applicable across different applications and industrial sectors, providing a basis for sensor performance comparison. PFS have several figures of merit that characterize their performance and utility, some of them being shown in Table 1.5.

The importance of each FOM depends on the specific application of the PFS. For example, a sensor used in wearable health monitoring may prioritize flexibility, comfort, and accuracy, while a sensor used in industrial automation may prioritize durability, accuracy, and response time.

Table 1.4 Main applications of printed flexible sensors

Sector	Application	Sensing Examples
Wearables and interactive textiles	Smart textiles	Clothing to monitor body posture, movements, and muscle activity for health monitoring and interactive interfaces
	Human health monitoring	To monitor physiological and vital signs (e.g., heart rate, blood pressure, skin moisture); integrated into fabrics or directly onto the skin
	Fitness tracking and sports monitoring	To track physical activity, performance, recovery (e.g., distance traveled, calories burned, sweat levels, muscle activity)
	Fashion and entertainment	Interactive clothing with LED and sensors
	Worker tracking and safety devices	Safety devices to detect falls, impacts, abnormal physiological condition, and trigger alerts
Medicine and healthcare	Medical diagnostics	Devices for rapid, non-invasive testing of biomarkers (e.g., glucose, ph, electrolytes, diabetes, and dehydration)
	Wearable health monitoring	Patches for real-time monitoring of vital signs (as above); monitoring patches (e.g., thin, pfs adhering to skin)
	Prosthetics and assistive devices	Feedback on movement and pressure in prosthetics and assistive devices (e.g., prosthesis, exoskeletons)
	Several medical devices	Temperature and humidity monitoring in medical patches and bandages. Pressure-sensitive mats; monitoring surgical tools
	Assisted living	Health monitoring, fall detection, environmental sensing, home automation, and enhancing the quality of life for the elderly and individuals with disabilities
Mobility and transports	Automotive	Monitoring vehicle performance, driver safety, vehicle condition, and safety; in-vehicle monitoring; connected vehicles (V2X)
	Aerospace and defense	Structural health monitoring (SHM) of commercial and military vehicles; data on structural fatigue, aerodynamic performance, and mission readiness for maintenance and mission planning
	Smart cities and mobility	IoT in smart environments (e.g., environmental monitoring, smart lighting); smart buildings (SHM, energy management, security); Intelligent transportation; smart parking; smart waste
Industry and robotics	Structural health monitoring (SHM)	Monitoring the structural integrity of assets (early warnings, predictive maintenance, repairs)

(continued)

Table 1.4 (continued)

Sector	Application	Sensing Examples
	CBM/PHM	Equipment health monitoring (condition-based monitoring, CBM) and predictive maintenance (PHM- prognostics and health management)
	Energy efficiency	Monitoring energy consumption and optimizing efficiency (buildings and industrial facilities)
	Asset tracking and inventory management	RFID tags for real-time visibility in supply chains; asset tracking, inventory management, improving operational efficiency
	Robotics and soft robotics	e-skin; tactile, proprioceptive sensing; environment, vibrations, impacts
	Food quality and safety	Monitoring food quality, freshness, and safety parameters; smart packaging, food safety; supply chain monitoring of perishable goods
	Agricultural	Monitoring soil moisture, nutrients, and crop health; precision agriculture techniques
Environment	Environmental monitoring	Detecting pollutants and monitoring air and water quality
	Air and water quality	Detecting pollutants/contaminants and air (e.g., particle sizes, toxins, CO_2, NO_2, O_3) and water quality (e.g., pH, contaminants)
	Gas sensing	Detecting and monitoring gases (CO, CO_2, CH_4, VOC) in various environments (industrial facilities, automotive, air quality)
Constructions	Structural health monitoring (SHM)	Monitoring the structural integrity of buildings, bridges, infrastructures (early warnings and predictive maintenance)
	Environmental monitoring	Measuring temperature, humidity, and pressure changes within structures; detecting moisture infiltration in walls, roofs, or foundations to prevent damage
	Worker and equipment safety	Monitoring equipment usage to ensure proper operation and prevent failures; detecting unsafe vibrations or movements in cranes, excavators, and other machinery; tracking worker health and environmental conditions (e.g., heat stress) on construction sites
	Construction site monitoring and automation	Autonomous construction equipment for improved precision and feedback; monitoring site conditions to optimize resource allocation and workflow; detecting unauthorized activity or equipment misuse

(continued)

Table 1.4 (continued)

Sector	Application	Sensing Examples
	Smart buildings and IoT integration	Embedding sensors in smart buildings for continuous environmental and structural monitoring; integrating PFS with IoT platforms for centralized data analysis and control; providing feedback for building management systems to optimize operations
Home and domestics	Smart home devices and appliances	PFS embedded in household appliances, smart home devices, temperature/humidity control appliances; personalized climate control
	Smart lightning	Automated brightness adjustment based on ambient light levels; controlling lighting based on occupancy detection; advanced lighting systems for circadian rhythm management
	Smart furniture and surfaces	Pressure sensors in furniture to monitor usage patterns and adjust ergonomics; embedded sensors in countertops or tables for touch-sensitive controls
	Assistive technologies	Motion sensors for fall detection and monitoring of elderly or disabled individuals; gesture recognition for controlling smart home devices without physical interaction; health monitoring sensors integrated into beds or chairs for continuous care
	Security and safety	Detection of unauthorized entry, glass breakage, or other signs of intrusion; monitor foot traffic; smoke and gas sensors for early fire or leak detection
Interactions devices	Human-computer (HCI) interactions	Touch-sensitive interfaces and haptic feedback; virtual reality and augmented reality applications
	Pressure-sensitive surfaces, tactile sensing	Pressure-sensitive keyboards, gaming controllers with force feedback for immersive experiences, smart surfaces that adjust interaction based on applied force, robotic grippers, and prosthetics for enhanced dexterity and sensitivity of robotic systems and assistive devices
	Smart textiles and wearable HMI	Clothing with embedded sensors for controlling devices through touch or motion, smart gloves for controlling drones or robotic systems, interactive shoes that adjust fit or provide input for sports analysis
	Interactive spaces	Touch-sensitive walls, tables, floors for immersive interactions; gesture-controlled interfaces in smart homes, offices, public buildings; smart surfaces for collaborative workspaces, entertainment

(continued)

Table 1.4 (continued)

Sector	Application	Sensing Examples
	Adaptive feedback systems	Haptic feedback for gaming, VR controllers, or prosthetics; smart chairs or surfaces that provide tactile cues for interaction; wearables that vibrate or change temperature based on device status
	Artificial skin (e-skin)	Mimicking human skin properties for realistic haptic feedback and sensory perception in interactive systems
	Smart packaging	Monitoring product freshness and tampering throughout the supply chain, ensuring product quality and safety
	Security, authentication	Biometric systems for secure access control

1.5 Evolution of Printed Flexible Sensor Technology

Initially developed for niche applications, PFS have evolved through the integration of advanced materials, improved printing techniques, and sophisticated design methodologies. The evolution of PFS technology can be traced back to the early development of electronic printing techniques in the mid-twentieth century (Fig. 1.3).

Initially, printed electronics were limited to simple conductive patterns using screen printing methods, primarily for basic circuitry and low-resolution applications. The 1980–90s saw significant advancements with the introduction of new materials, such as conductive polymers and organic semiconductors, which allowed for greater flexibility and functionality of the printed devices. The turn of the millennium marked a pivotal point with the advent of inkjet printing technology, enabling high-resolution patterning and the ability to print on various flexible substrates. This period also saw the development of novel nanomaterials, such as carbon nanotubes (CNT) and graphene, which brought unprecedented electrical properties and mechanical strength to the printed sensors. The 2010s witnessed an increment in research and commercial interest in PFS, driven by the growing demand for wearable technology, environmental monitoring, and smart devices. Roll-to-roll manufacturing processes emerged during this time, facilitating large-scale production, reducing costs, and making these sensors more accessible for widespread use. Throughout the past decade, the integration of advanced materials science and printing technologies has continued to evolve. Innovations such as new functional and superior inks, 3D printing and hybrid printing techniques have further expanded the capabilities of PFS. This era also saw the development of multifunctional and hybrid sensors capable of detecting various physical, chemical, and biological parameters simultaneously. In

Table 1.5 Main figures of merit of printed flexible sensors

Figure of Merit (FOM)	Description	Examples
Flexibility	This flexibility FOM (FFOM) assesses how well a flexible sensor conforms to arbitrary surfaces, how it bends, stretches, deforms without breaking, and loses its functionality	ε_y/t (yield strain/thickness), radius of curvature (ρ): $FFOM = \frac{1}{\rho}\frac{1}{E}\frac{1}{\Delta Elec.Prop.}$ (E—Young's modulus)
Sensitivity	Measures the change in output signal per unit change in the input parameter being measured; higher sensitivity means that the sensor can detect smaller changes in the parameter	Ω/με, Ω/°C, V/% (Ω—electrical resistivity, με—microstrain, °C—° Celsius, V—voltage)
Durability	Measures how well the sensor withstands repeated bending or loading without degradation in performance; a durable sensor shall have a longer operational lifetime	N° bending cycles
Response time	How quickly the sensor reacts to changes in the variable being measured; faster response times contribute positively to sensitivity in dynamic environments	Time taken reach, e.g., 90% of final value due to changes in the parameter
Linearity	Describes how well the sensor's output follows a linear relationship with changes in the input parameter; a linear sensor simplifies calibration and data interpretation	R^2, % FS (R^2—coefficient of linear correlation, FS—full-scale)
Accuracy	How closely the sensor's measurements match the true value of the input parameter being measured; high accuracy ensures reliable data output	% or absolute error
Repeatability	Ability of the sensor to produce consistent measurements under the same input and conditions, multiple times	Standard deviation, variance
Resolution	The smallest detectable change in tx\he input parameter that the sensor can detect reliably. Higher resolution allows for more precise measurements; lower LOD implies higher sensitivity, as the sensor can detect very minimal changes; high SNR means the signal changes are distinct from background noise, enhancing sensitivity	Smallest detectable change in input, limit of detection (LOD), signal-to-noise ratio (SNR)
Hysteresis	Difference in output readings when the input parameter is changed in one direction compared to when it is changed in the opposite one; low hysteresis indicates good repeatability and consistency of measurements	Hysteresis loop area, max. difference ascending and descending curves at given point, % FS output

(continued)

Table 1.5 (continued)

Figure of Merit (FOM)	Description	Examples
Drift	Gradual change in the sensor's output over time (e.g., hours) when exposed to a constant input or operating condition; this long-term instability affects the accuracy and reliability of the measurements	°C/h, °/h
Operating range	Is the minimum and maximum temperatures over which the sensor can reliably operate; it is important to ensure that the sensor is suitable for the intended application's temperature range	[min, max]
Temperature stability or range	How well the sensor maintains its performance across a range of temperatures; temperature fluctuations can affect sensor readings, so stability is important (e.g., outdoor or harsh environments)	Change in output/°C
Cost	Overall cost of the sensor, including manufacturing, installation, maintenance; lower-cost sensors are more suitable for mass deployment or disposable applications	Total cost (materials, manufacturing, operation, maintenance)
Selectivity/ Specificity	Ability to distinguish the target analyte from other interfering substances present; high selectivity ensures accurate and reliable detection of the target analyte without interference from other components	Specific analyte
Regeneration capability	Gas/chemical sensor's or biosensor's ability to be reused multiple times by restoring its sensing capability after each measurement cycle	Signal recovery, changes on sensitivity, selectivity, and response time

recent years, the focus shifted toward enhancing performance, reliability, and integration of PFS in complex systems. This includes the incorporation of energy-harvesting elements, advanced signal processing, and wireless communication capabilities to create fully autonomous sensing systems.

The economic and social impact of PFS technology is profound, driving innovation and efficiency across multiple sectors. Economically, the global market for PFS is expected to reach \$8.6 billion by 2031, growing at a compound annual growth rate (CAGR) of 8.3%. These sensors have enabled cost-effective production methods

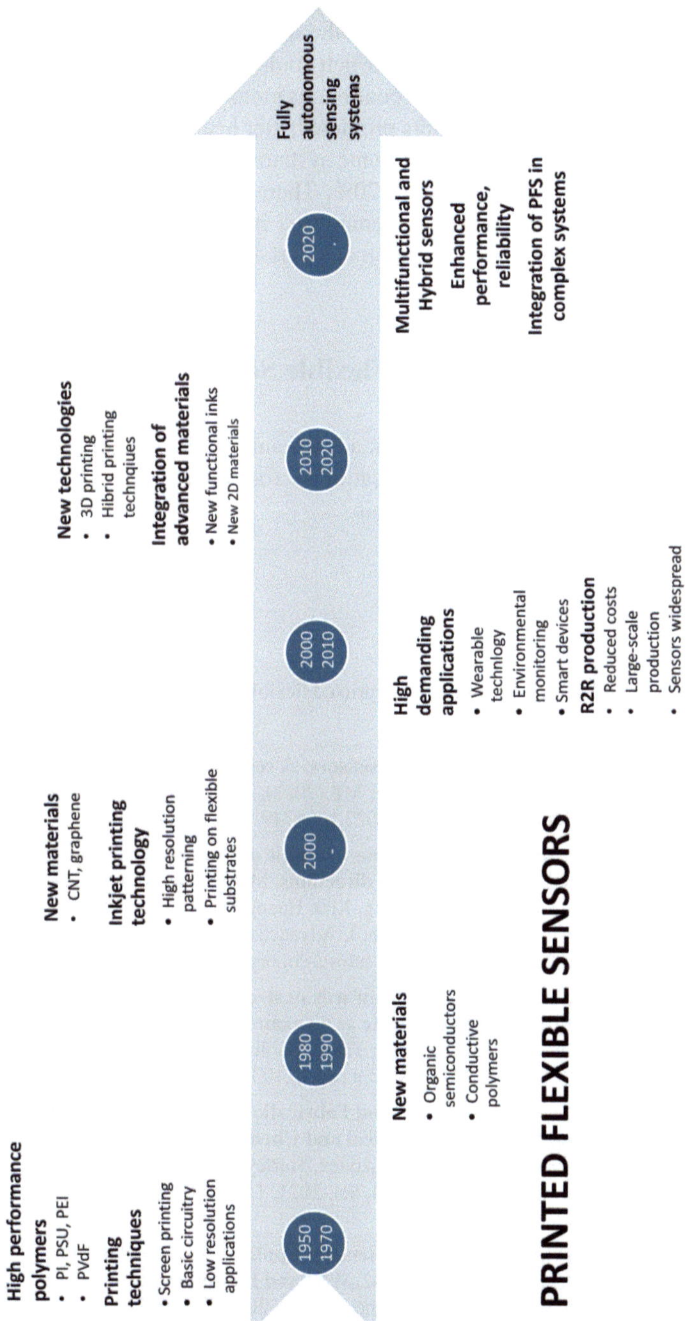

Fig. 1.3 Historic chronogram of the evolution of PFS technology

and scalable manufacturing, reducing costs for advanced sensing technologies and making them accessible to a broader range of industries.

Socially, PFS have significantly improved the quality of life by facilitating the development of wearable health monitors, which could help reduce healthcare costs by up to 30% through early diagnosis and continuous monitoring. Additionally, smart home devices and environmental sensors promote better health, safety, and environmental management. For instance, smart home systems incorporating these sensors can reduce energy consumption by up to 20%. Their ability to provide real-time, accurate data enhances decision-making, empowers individuals with better health management tools, and supports the creation of smart and sustainable communities.

1.6 Review Articles on Printed Flexible Sensors

Recent reviews on PFS highlight significant advancements in this field, focusing on various printing techniques and their applications across industries.

Table 1.6 lists relevant works on this topic.

Table 1.6 Literature review articles published on printed flexible sensors

Topic	References
Materials, inks, substrates	**MXene-based flexible sensors: A review,** Hajian S, Maddipatla D, Narakathu BB, Atashbar MZ (2022), Front. Sens. 3:1006749, https://doi.org/10.3389/fsens.2022.1006749
	Graphene-based flexible wearable sensors: mechanisms, challenges, and future directions, Ming Kong, Min Yang, Runze Li, Yun-Ze Long, Jun Zhang, Xian Huang, Xin Cui, Yanbin Zhang, Zafar Said, Changhe Li, Intern. J. Advanced Manufacturing Technology (2024) 131:3205–3237, https://doi.org/10.1007/s00170-023-12007-7
	Printable biosensors towards next-generation point-of-care testing: paper substrate as an example, Yaolin Liu, Sheng Lu, Zhiheng Zhang, Zhugen Yang, Xiaolin Cui, Guozhen Liu, Lab Chip, 2023, 23, 3328, https://doi.org/10.1039/d3lc00038a
	Review of Materials and Fabrication Methods for Flexible Nano and Micro-Scale Physical and Chemical Property Sensors, Anesu Nyabadza, Mercedes Vázquez, Shirley Coyle, Brian Fitzpatrick, Dermot Brabazon, Appl. Sci. 2021, 11, 8563. https://doi.org/10.3390/app11188563
	2D Materials for Gas Sensing Applications: A Review on Graphene Oxide, MoS2, WS2 and Phosphorene, Maurizio Donarelli, Luca Ottaviano, Sensors 2018, 18, 3638; https://doi.org/10.3390/s18113638

(continued)

Table 1.6 (continued)

Topic	References
	Flexible Graphene-Based Wearable Gas and Chemical Sensors, Eric Singh, M. Meyyappan, Hari Singh Nalwa, ACS Appl. Mater. Interfaces 2017, 9, 34544−34586, https://doi.org/10.1021/acsami.7b07063
	Flexible Field-Effect Transistor-Type Sensors Based on Conjugated Molecules, Yoon Ho Lee, Moonjeong Jang, Moo Yeol Lee, O. Young Kweon, Joon Hak Oh, Chem 3, 724–763, November 9, 2017, https://doi.org/10.1016/j.chempr.2017.10.005
	A Review of Carbon-Based Conductive Inks and Their Printing Technologies for Integrated Circuits, Yufeng Qin, Xueqiong Ouyang, Yang Lv, Wencai Liu, Qing Liu, Shuangxi Wang, Coatings 2023, 13, 1769. https://doi.org/10.3390/coatings13101769
	Semiconductors for organic transistors, Antonio Facchetti, Materials Today, 2007, 10 (3), https://doi.org/10.1016/S1369-7021(07)70017-2
	Flexible Field-Effect Transistor-Type Sensors Based on Conjugated Molecules, Yoon Ho Lee, Moonjeong Jang, Moo Yeol Lee, O. Young Kweon, Joon Hak Oh, Chem 3, 2017, 724–763, https://doi.org/10.1016/j.chempr.2017.10.005
	PVDF-based dielectric polymers and their applications in electronic materials, Weimin Xia, Zhicheng Zhang, IET Nanodielectr., 2018, Vol. 1 Iss. 1, pp. 17-31, https://doi.org/10.1049/iet-nde.2018.0001
	Chemistry of solid metal-based inks and pastes for printed electronics—A review, Clara Cano-Raya, Zlatan Z. Denchev, Sílvia F. Cruz, Júlio C. Viana, Applied Materials Today, 15, 2019, pp. 416-430, https://doi.org/10.1016/j.apmt.2019.02.012
Fabrication techniques and integration	**Electrode manufacturing based on printing: a mini review,** Hao Zheng, Zijing Guo, Wangwang Zhu, Dachao Li, Zhihua Pu, The International Journal of Advanced Manufacturing Technology (2023) 128:2813–2824, https://doi.org/10.1007/s00170-023-12115-4
	A review of connectors and joining technologies for electronic textiles, Jessica Stanley, John A. Hunt, Phil Kunovski4 YangWei, Engineering Reports. 2022;4:e12491, https://doi.org/10.1002/eng2.12491
	Recent Advancements in Liquid Metal Flexible Printed Electronics: Properties, Technologies, and Applications, Xuelin Wang, Jing Liu, Micromachines 2016, 7, 206; https://doi.org/10.3390/mi7120206
	A Review on Printed Electronics with Digital 3D Printing: Fabrication Techniques, Materials, Challenges and Future Opportunities, C. Hanumanth Rao, Kothuru Avinash, B. K. S. V. L. Varaprasad, Sanket Goel, Journal of Electronic Materials (2022) 51:2747–2765, https://doi.org/10.1007/s11664-022-09579-7

(continued)

Table 1.6 (continued)

Topic	References
	3D Printed Electronics of Non-contact Ink Writing Techniques: Status and Promise, Haining Zhang, Seung Ki Moon, Teck Hui Ngo, International Journal of Precision Engineering and Manufacturing-Green Technology (2020) 7:511–524 https://doi.org/10.1007/s40684-019-00139-9
	Laser-Assisted Printed Flexible Sensors: A Review, Tao Han, Anindya Nag, Nasrin Afsarimanesh, Subhas Chandra Mukhopadhyay, Sudip Kundu, Yongzhao Xu, Sensors 2019, 19, 1462; https://doi.org/10.3390/s19061462
	Recent Progress in Manufacturing Techniques of Printed and Flexible Sensors: A Review, Dinesh Maddipatla, Binu B. Narakathu, Massood Atashbar, Biosensors 2020, 10, 199; https://doi.org/10.3390/bios10120199
	Inkjet-printed flexible sensors: From function materials, manufacture process, and applications perspective, Xuepei Wang, Mengjie Zhang, Luwei Zhang, Jinchao Xu, Xinqing Xiao, Xiaoshuan Zhang, Materials Today Communications 31 (2022) 103263, https://doi.org/10.1016/j.mtcomm.2022.103263
	Printed Strain Sensors for Motion Recognition: A Review of Materials, Fabrication Methods, and Machine Learning Algorithms, Nathan Zavanelli, Kangkyu Kwon, Woon-Hong Yeo, J. Eng. Medicine and Biology. 2023, 1-35, https://doi.org/10.1109/OJEMB.2023.3330290
	Technology Roadmap for Flexible Sensors, Yifei Luo, et.al., ACS Nano 2023, 17, 5211−5295, https://doi.org/10.1021/acsnano.2c12606
	3D Printing of Flexible Electronic Devices, Hui Yang, Wan Ru Leow, and Xiaodong Chen, Small Methods 2018, 2, 1700259, https://doi.org/10.1002/smtd.201700259
	3D printed electronics: Processes, materials and future trends, Hong Wei Tan, Yu Ying Clarrisa Choong, Che Nan Kuo, Hong Yee Low, Chee Kai Chua, Progress in Materials Science 127 (2022) 100945, https://doi.org/10.1016/j.pmatsci.2022.100945
	Recent Advances in Flexible Piezoresistive Arrays: Materials, Design, and Applications, Shuoyan Xu, Zigan Xu, Ding Li, Tianrui Cui, Xin Li, Yi Yang, Houfang Liu, Tianling Ren, Polymers 2023, 15, 2699. https://doi.org/10.3390/polym15122699
Designs	**Review of Printed Electrodes for Flexible Devices,** Luo Z, Qiao F, Wu X and Zhang J (2019) Front. Mater. 5:77, https://doi.org/10.3389/fmats.2018.00077
	Flexible electronics under strain: a review of mechanical characterization and durability enhancement strategies, K. D. Harris, A. L. Elias, H.-J. Chung, J Mater Sci (2016) 51:2771–2805, https://doi.org/10.1007/s10853-015-9643-3
	Review of Printed Electrodes for Flexible Devices, Li Q, Zhang J, Li Q, Li G, Tian X, Luo Z, Qiao F, Wu X and Zhang J, (2019) Front. Mater. 5:77, https://doi.org/10.3389/fmats.2018.00077

(continued)

Table 1.6 (continued)

Topic	References
	Theory, technology and applications of piezoresistive sensors: A review, A.S. Fiorillo, C.D. Critello, S.A. Pullano, Sensors and Actuators A 281 (2018) 156–175, https://doi.org/10.1016/j.sna.2018.07.006
	Strategies for Designing Stretchable Strain Sensors and Conductors, Shuying Wu, Shuhua Peng, Yuyan Yu, Chun-Hui Wang, Adv. Mat. Techn., 5 (2), 2020, 1900908, First published: 15 January 2020 https://doi.org/10.1002/admt.201900908
	Recent advances in inkjet-printing technologies for flexible/wearable electronics, Yu Liu, Hongze Zhu, Lei Xing, Qingkai Bu, Dayong Ren, Bin Sun, Nanoscale, 2023, 15, 6025–6051. https://doi.org/10.1039/d2nr05649f
	The design, fabrication, and applications of flexible biosensing devices, Meng Xu, Dora Obodo, Vamsi K. Yadavalli, Biosensors and Bioelectronics 124–125 (2019) 96–114, https://doi.org/10.1016/j.bios.2018.10.019
	Flexible pressure sensors with microstructures, Ruitao Tang, Fangyuan Lu, Lanlan Liu, Yu Yan, Qifeng Du, Bocheng Zhang, Tao Zhou, Haoran Fu, Nano Select 2021; 2:1874–1901, https://doi.org/10.1002/nano.202100003
	A Review of Vertical Organic Transistors, Hans Kleemann, Kevin Krechan, Axel Fischer, Karl Leo, Adv. Funct. Mater. 2020, 30, 1907113, https://doi.org/10.1002/adfm.201907113
	Crack engineering boosts the performance of flexible Sensors, Yunlei Zhou, Haoxiang Lian, Zhenlei Li, Liting Yin, Qian Ji, Kan Li, Fei Qi, YongAn Huang, VIEW, 2022; 3:20220025, https://doi.org/10.1002/VIW.20220025
PFS and applications	**Flexible Electronics: Status, Challenges and Opportunities,** Daniel Corzo, Guillermo Tostado-Blázquez, Derya Baran, Front. Electron. 1:594003, https://doi.org/10.3389/felec.2020.594003
	Recent progress in printing flexible electronics: A review, BI Sheng, GAO BuHan, HAN Xu, HE ZhengRan, METTS Jacob, JIANG ChengMing, ASARE-YEBOAH Kyeiwaa, Sci China Tech Sci, 2023, 66, https://doi.org/10.1007/s11431-021-2093-4
	A Review on Printed Electronics: Fabrication Methods, Inks, Substrates, Applications and Environmental Impacts, Wiklund, J.; Karakoç, A.; Palko, T.; Yi ˘ gitler, H.; Ruttik, K.; Jäntti, R.; Paltakari, J., J. Manuf. Mater. Process., 2021, 5, 89. https://doi.org/10.3390/jmmp5030089
	Recent Advances in the Development of Flexible Sensors: Mechanisms, Materials, Performance Optimization, and Applications, Wenguang Yang, Yitong Qin, Zhen Wang, Tao Yu, Zhixing Ge, Journal of Electronic Materials (2022) 51:6735–6769, https://doi.org/10.1007/s11664-022-09922-y

(continued)

Table 1.6 (continued)

Topic	References
	A New Frontier of Printed Electronics: Flexible Hybrid Electronics, Y. Khan, A. Thielens, J. Ting, C. Baumbauer, A. C. Arias, Adv. Mater. 2020, 32, 190527, https://doi.org/10.1002/adma.201905279
	Printable, Highly Sensitive Flexible Temperature Sensors for Human Body Temperature Monitoring: A Review, Yi Su, Chunsheng Ma, Jing Chen, Huiping Wu, Weixiang Luo, Yueming Peng, Zebang Luo, Lin Li, Yongsong Tan, Olatunji Mumini Omisore, Zhengfang Zhu, Lei Wang, Hui Li, Nanoscale Res Lett (2020) 15:200, https://doi.org/10.1186/s11671-020-03428-4
	Theory, technology and applications of piezoresistive sensors: A review, A.S. Fiorillo, C.D. Critello, S.A. Pullano, Sensors and Actuators A 281 (2018) 156–175, https://doi.org/10.1016/j.sna.2018.07.006
	Flexible Electronics: Status, Challenges and Opportunities, Daniel Corzo, Guillermo Tostado-Blázquez, Derya Baran, Front. Electron. 1:594003, https://doi.org/10.3389/felec.2020.594003
	Printed flexible mechanical sensors, Samuel Smocot, Zixin Zhang, Lingzhi Zhang, Shu Guo, Changhong Cao, Nanoscale, 2022, 14, 17134, https://doi.org/10.1039/d2nr04015h
	Recent Advances in Flexible Sensors and Their Applications, Bouchaib Zazoum, Khalid Mujasam Batoo, Muhammad Azhar Ali Khan, Applications. Sensors 2022, 22, 4653, https://doi.org/10.3390/s22124653
	Recent Advances in the Development of Flexible Sensors: Mechanisms, Materials, Performance Optimization, and Applications, Wenguang Yang, Yitong Qin, Zhen Wang, Tao Yu, Zhixing Ge, Journal of Electronic Materials (2022) 51:6735–6769, https://doi.org/10.1007/s11664-022-09922-y
	Printed Strain Sensors for On-Skin Electronics, Yizong Li, Yuxuan Liu, Shah Rifat Alam Bhuiyan, Yong Zhu, Shanshan Yao, Small Struct. 2022, 3, 2100131 2100131, pp.1-23, https://doi.org/10.1002/sstr.202100131
	A review of flexible printed sensors for automotive infotainment Systems, K. P. Srinivasan, T. Muthuramalingam, Ammar H. Elsheikh, Archives of Civil and Mechanical Engineering (2023) 23:67, https://doi.org/10.1007/s43452-023-00604-y
	Recent progress on screen-printed flexible sensors for human health monitoring, Xue Gong, Kai Huang, Yu-Hang Wu, Xiao-Sheng Zhang, Sensors & Actuators: A. Physical 345 (2022) 113821, https://doi.org/10.1016/j.sna.2022.113821
	Recent advances in wearable tactile sensors: Materials, sensing mechanisms, and device performance, Tingting Yanga, Dan Xie, Zhihong Lib, Hongwei Zhua, Materials Science and Engineering R 115 (2017) 1–37, http://dx.doi.org/10.1016/j.mser.2017.02.001

(continued)

Table 1.6 (continued)

Topic	References
	Advanced flexible humidity sensors: structures, techniques, mechanisms and performances, Yuzhe Zhang, Yuxi Liu, Lifei Lin, Man Zhou, Wang Zhang, Liwei L., Front. Mater. Sci. 2023, 17(4): 230662, https://doi.org/10.1007/s11706-023-0662-8
	The Progress of Research into Flexible Sensors in the Field of Smart Wearables, Yunlei Yin, Cheng Guo, Hong Li, Hongying Yang, Fan Xiong,Dongyi Chen, Sensors 2022, 22, 5089, https://doi.org/10.3390/s22145089
	Flexible, Stretchable Sensors for Wearable Health Monitoring: Sensing Mechanisms, Materials, Fabrication Strategies and Features, Yan Liu, Hai Wang, Wei Zhao, Min Zhang, Hongbo Qin, Yongqiang Xie, Sensors 2018, 18, 645; https://doi.org/10.3390/s18020645
	Flexible/wearable resistive gas sensors based on 2D materials, Hamid Reza Ansari, Ali Mirzaei, Hooman Shokrollahi, Rahul Kumar, Jin-Young Kim, Hyoun Woo Kim, Mahesh Kumar, Sang Sub Kim, J. Mater. Chem. C, 2023, 11, 6528, https://doi.org/10.1039/d3tc00806a
	Recent Advances in Flexible Field-Effect Transistors toward Wearable Sensors, Ming-Zheng Li, Su-Ting Han, Ye Zhou, Adv. Intell. Syst. 2020, 2, 2000113, https://doi.org/10.1002/aisy.202000113
	Organic field-effect transistor-based flexible sensors, Saravanan Yuvaraja, Ali Nawaz, Qian Liu, Deepak Dubal, Sandeep G. Surya, Khaled N. Salama, Prashant Sonar, Chem. Soc. Rev., 2020, 49, 3423, https://doi.org/10.1039/c9cs00811j
	Advances in flexible organic field-effect transistors and their applications for flexible electronics, Kai Liu, Bang Ouyang, Xiaojun Guo, Yunlong Guo, Yunqi Liu, npj Flexible Electronics, 2022, 6:1; https://doi.org/10.1038/s41528-022-00133-3
	Recent progress in printing flexible electronics: A review, Bi S, Gao B H, Han X, et al.. Sci China Tech Sci, 2023, 66, https://doi.org/10.1007/s11431-021-2093-4
	Review—An Overview on Recent Progress in Screen-Printed, Electroanalytical (Bio)Sensors, Sima Singh, Joseph Wang, Stefano Cinti, ECS Sensors Plus, 2022 1 023401, https://doi.org/10.1149/2754-2726/ac70e2
	Organic field-effect transistor-based flexible Sensors, Saravanan Yuvaraja, Ali Nawaz, Qian Liu, Deepak Dubal, Sandeep G. Surya, Khaled N. Salama, Prashant Sonar, Chem. Soc. Rev., 2020, 49, 3423, 10 https://doi.org/.1039/c9cs00811j
	SAW Sensors for Chemical Vapors and Gases, Jagannath Devkota, Paul R. Ohodnicki, David W. Greve, Sensors 2017, 17, 801; https://doi.org/10.3390/s17040801

Chapter 2
Fundamentals of Printed Flexible Sensors

The fundamentals of PFS encompass several key aspects, including materials (flexible substrates, functional inks), fabrication techniques (pre-printing, printing, post-printing, additive manufacturing), and integration of PFS with electronics. The economic and social impact of PFS technology is profound, driving innovation across multiple sectors. Economically, the global market for PFS is estimated to reach $8.6 billion by 2031, growing at a compound annual growth rate of 8.3% [3]. These sensors have cost-effective production methods and scalable manufacturing, reducing costs for advanced sensing technologies and making them accessible to a broader range of industries. Socially, PFS have significantly improved the quality of life by facilitating the development of wearable for health monitoring, reducing healthcare costs through early diagnosis and continuous monitoring. Additionally, smart home devices and environmental sensors promote better health, safety, and environmental management. For instance, smart home systems incorporating these sensors can reduce energy consumption. Their ability to provide real-time and accurate data enhances decision-making, empowers individuals with better health management tools, and supports the creation of smart, sustainable communities.

2.1 Substrates for Flexibility

Flexible substrates provide mechanical support, flexibility, and compatibility with printing processes [2]. Some commonly flexible substrates used for printed electronics are presented in Table 2.1.

Polymeric substrates are crucial in the development of flexible electronics and printed sensors. These substrates, made from various types of polymers, offer excellent flexibility, mechanical strength, lightweight, ease of processing, durability, thermal stability, and often biocompatibility and transparency. Polymers offer a range of suitable properties that can be tuned to the application (e.g., biocompatible,

J. C. Viana, *Primer on Printed Flexible Sensors*,
https://doi.org/10.1007/978-3-032-06554-4_2

Table 2.1 Main substrate materials for printed flexible sensors

Flexible substrate	Materials	Properties
Polymers	Thermoplastics: polyimide (PI), polyethylene terephthalate (PET), polyethylene naphthalate (PEN), polycarbonate (PC), thermoplastic elastomer (TPE), recycled polymers, PVdF, biodegradable polymers (PLA, PHA, starch-based), others	Can be melted and reshaped multiple times, can be transparent, are flexible, lightweight, and ease of processing
	Thermosets: epoxy resins, polyurethanes, phenolic resins, melamine resins, polyester resins, etc.	Undergo irreversible chemical cross-linking during curing; good mechanical properties and chemical resistance; low flexibility
	Elastomers: polydimethylsiloxane (PDMS), Silicone rubber, Natural rubber (NR), Nitrile rubber (NBR), Styrene-butadiene rubber (SBR), etc.	Offer flexibility, stretchability, and high resilience; can be transparent
Paper	Photopaper, Cellulose, Kraft paper, recycled paper, etc.	Biodegradable; recyclable, lightweight, flexible, environmentally friendly
Textiles	Textiles fabrics and non-fabrics, cotton, polyester, nylon, others	Breathable; durable, stretchable, flexible
Thin metallic foils	Al, Cu, stainless steel	High electrical and thermal conductivity, flexible, reflective surfaces
Thin flexible glass	Flexible glass	High transparency; scratch resistance, flexible but with limited bend radius, excellent barrier properties
Human skin	–	Stretchable, self-healing capabilities

biodegradable). Polymer substrates can be classified into three main types: thermoplastics, thermosets, and elastomers. **Thermoplastic** polymers are flexible, and easy moldable, being able to be reshaped and processed multiple times without significant degradation. **Thermoset** polymers irreversibly cure to form a hard, infusible, and insoluble material. They undergo a chemical reaction that creates cross-linking between polymer chains, resulting in a three-dimensional rigid network. **Elastomers** offer high flexibility, stretchability, and high resilience but, generally, they have low Young's modulus and a high failure strain, making them suitable for flexible sensor applications that require high mechanical deformability.

Paper, although traditionally associated with printing text and images, has evolved into an essential substrate for various electronic applications (e.g., RFID and NFC tags, smart packaging) and sensors (e.g., biosensors, disposable sensors). It is flexible, lightweight, low-cost, recyclable, biodegradable, and printable, being an attractive choice for a wide range of disposable electronics and sensor devices.

Textiles, such as fabrics and textiles, have become increasingly important substrates for electronic and sensor applications. The integration of electronics into textiles has led to the development of smart textiles (e-textiles), which offer functionalities such as sensing, actuation, communication, and energy harvesting. Textile substrates offer comfort, breathability, and conformability to the user's body, allowing for the integration of sensors directly into clothing, accessories, or wearable devices.

Thin metal foils can serve as flexible substrates for printed sensors, particularly in applications that require high electrical conductivity and mechanical robustness. Metal foils offer excellent electrical properties and thermal conductivity, making them suitable for flexible sensor applications in harsh environments or high-temperature conditions.

Flexible glass substrates combine the high-performance characteristics of traditional glass with some flexibility. Key properties are high transparency, flexibility, scratch resistance, excellent barrier properties against humidity and gases, excellent thermal and chemical resistance, and durability, maintaining device integrity in various environments.

Human skin serves as a natural substrate for various electronic and sensor applications, particularly in the field of biomedical devices. By integrating electronics and sensors directly onto the skin (e-tattoos), it allows to create non-invasive, comfortable, and conformable devices for healthcare monitoring, human-machine interfaces, and augmented reality applications.

Table 2.2 lists the main properties of flexible substrates for PFS.

The selection of the substrate material for PFS depends upon factors such as the application requirements, the required mechanical properties, the involved sensing mechanism, the compatibility with the printing processes, and other functional materials. Some common requirements are:

- flexibility—that depends upon the substrate thickness and material Young's modulus;
- stretchability—maximum allowable deformation (e.g., yield strain or strain at break);
- mechanical durability—mainly during repeated bending;
- chemical resistance—exposure to solvents, humidity, and environmental contaminants;
- dimensional stability—exhibit minimal dimensional changes (e.g., expansion or shrinkage over a wide range of temperatures and humidity levels; low coefficient of thermal expansion, CTE);
- smooth surface—for achieving uniform deposition and adhesion of printed materials and better sensor performance;
- electrical insulation—to prevent short circuits, minimize leakage currents, and interference between sensor elements (a dielectric coating can be applied);

Table 2.2 Main properties of flexible substrates for printed flexible sensors

Material	Density	CTE (%.C^{-1})	K (W/ m.K)	Thermal stability (°C)	Modulus (GPa)	Extensibility (%)	Transparency
PET	1.38	70–100	0.15–0.24	<80	2.1–2.4	50–150	Yes
PEN	1.37	20–30	0.15–0.27	<160	5.5–6.0	100–150	Yes
PC	1,20	65–70	0.19–0.22	<140	2.0–2.4	100–150	Yes
PI	1.43	20–40	0.12–0.14	<350	2.5–3.0	10–50	Yes, orange
PDMS	0,97	310–400	0.15	<180	0.36–0.87	150–200	Yes
TPU	1.1–1.3	90–160	0.21–0.28	<100	0.05–2.25	300–700	No
Photopaper	0.7–1.2	50–100	0.05–0.10	<100	1–2	2–5	No
Textile fabric	0.9–1.5	50–100	0.04–0.06	<250	5–10	20–40	No
Textile non-fabric	0.9–1.5	50–100	0.04–0.06	<250	5–10	20–40	No
Thin metallic foils	2.7–8.9	2–25	100–400	<1200	70–200	1–5	No
Thin flexible glass	2.5	3–8	1.0–1.2	<600	60–90	<1	Yes
Human skin	1.09	150–350	0.2–0.5	<45	0.001–0.3	35.65	No

- thermal stability—to withstand the temperatures encountered during printing processes and subsequent device operation;
- compatibility with printing techniques—allowing for precise patterning and deposition of sensor materials, with appropriate viscosity, surface energy, and wettability to facilitate ink adhesion and spreading;
- cost-effectiveness—readily available in large quantities to enable scalable manufacturing of PFS.

By fulfilling these requirements, the flexible substrate provides a solid foundation for the fabrication and performance of PFS, ensuring their functionality, reliability, and suitability for various applications.

2.2 Functional Inks and Materials

Functional inks are specialized formulations of inks designed for printing electronic components and devices using techniques such as inkjet, screen, flexography, and gravure printing. These inks contain materials that exhibit specific electrical, optical, or chemical properties, allowing them to function as electrical conductors, semiconductors, insulators, or active layers in printed electronics and sensors [4]. Depending upon the sensing mechanism and application, various functional materials are used to impart sensitivity to specific stimuli, such as pressure, strain, temperature, humidity, gas, or biochemical analytes. There are several types of functional inks commonly used in printed electronics and sensors:

Conductive Inks: These inks contain materials such as metallic nanoparticles (e.g., Ag, Cu, Au) [5], carbon allotropes (e.g., CNT, graphene) [6], or conductive polymers (e.g., PEDOT: PSS) suspended in a solvent. They are used to print conductive traces, electrodes, interconnects in electronic circuits, and sensors.

Semiconductive Inks: These inks contain semiconducting materials such as a) organic semiconductors (e.g., P3HT and other organic molecules), which are used for creating flexible transistors and other semiconductor devices, offering good mechanical properties and can be processed at low temperatures; b) oxide semiconductors, which are inks containing materials like zinc oxide (ZnO); or c) quantum dots; semiconductor inks are used to print active layers in electronic devices such as transistors, diodes, and photodetectors.

Dielectric inks: These inks contain insulating materials such as polymers, ceramics, or dielectric nanoparticles; UV-curable dielectric inks are used to insulate and protect conductive traces. Flexible polymer dielectrics, such as polyurethane or silicone-based inks, provide good electrical insulation and mechanical flexibility. Dielectric inks are used to print insulating layers and capacitors in electronic circuits and sensors.

Piezoelectric inks: Piezoelectric inks are specialized inks that contain piezoelectric materials capable of generating electrical charges when deformed by a mechanical stress. These inks are used in printing technologies to create sensors and devices that exploit the piezoelectric effect. Main flexible inks are based on PVdF (polyvinylidene fluoride), or composite inks combining piezoelectric particles (e.g., PZT, $BaTiO_3$) with flexible polymers.

Magnetic Inks: are inks that include ferromagnetic or ferrimagnetic particles (e.g., iron oxide, cobalt). They are used in magnetic storage devices, anti-counterfeiting measures, and sensors.

Optical inks: are specialized inks that exhibit unique optical properties, such as fluorescence, phosphorescence, color-shifting, or other light-related behaviors. They contain compounds that emit light upon excitation (fluorescence) or after excitation (phosphorescence). Applications include security printing, bioimaging, and decorative applications. Color-shifting inks show a change color based on the angle of view or lighting conditions and are used for anti-counterfeiting and decorative purposes. Optical inks are used in various applications, including security printing, decorative purposes, and functional devices.

Chromogenic inks: are specialized inks that change color in response to specific external stimuli such as light (photochromic), temperature (thermochromic), or chemical reactions. These inks are used in various applications, including security printing, packaging, consumer products, and novelty items due to their unique color-changing properties.

Quantum Dot inks: are inks that contain semiconductor nanoparticles that exhibit size-dependent optical properties, emitting specific colors when excited by light. They have applications in high-resolution displays, LED lighting, and advanced bioimaging.

Functionalized inks: These inks contain materials functionalized with specific chemical groups or biomolecules for sensing applications. They are used to print sensing elements, such as antibodies, enzymes, or DNA probes, for detecting analytes in biological, chemical, or environmental samples.

Hybrid Inks: These inks combine multiple functional materials to achieve specific properties or functionalities. For example, hybrid inks may combine conductive and semiconductive materials to print electronic devices with integrated sensing capabilities.

Stretchable Inks: These inks are specialized conductive inks designed to maintain electrical conductivity while being stretched, bent, or deformed. They include liquid metal inks, such as gallium-based alloys, which maintain electrical conductivity even when stretched. They are used in applications requiring high elasticity, such as stretchable circuits, wearable sensors, and soft robotics. Elastomeric conductive inks are based on elastomers, like PDMS, polyurethane, thermoplastic elastomers, and other flexible polymers that are mixed with conductive fillers. They are used for highly stretchable and flexible sensor applications. Polymer-based conductive inks offer good stretchability and can be processed at relatively low temperatures. They consist of conductive polymers like PEDOT: PSS, often combined with stretchable polymers, being used in organic electronics, flexible displays, and biointegrated sensors. CNT and graphene-based inks are composed of SWCNT, MWCNT, graphene flakes, or graphene oxide dispersed in a stretchable medium. They combine high electrical conductivity with mechanical flexibility and strength, being used for flexible and wearable electronics, flexible transistors, stretchable sensors, and conductive textiles, and other applications. Stretchable inks are crucial for applications in wearable electronics, flexible displays, soft robotics, and other fields requiring flexible and durable electronic components. Table 2.3 compares the properties of commercially available electrically conductive stretchable inks.

Functional inks enable the rapid and cost-effective fabrication of PFS on a variety of substrates. They must meet several requirements to ensure performance, durability, and flexibility. These requirements encompass properties of electrical conductivity, adhesion, flexibility, and stability under various environmental and mechanical conditions. Main requirements for functional inks are:

- High electrical conductivity: Conductive inks should offer high electrical conductivity to ensure efficient current flow. This is especially crucial for applications in sensors, antennas, and interconnects. Inks should maintain consistent electrical performance in various mechanical deformations to enable reliable sensing and electronic operation. Some applications may require tunable resistivity (e.g., resistors and flexible heaters), allowing to meet specific resistance needs.
- Adhesion to flexible substrates, without delaminating or cracking during deformation. Some substrates may require surface treatments (e.g., plasma treatment)

Table 2.3 Electrically conductive stretchable inks for printing PFS

Ink type	Stretchability (%)	Electrical conductivity (S/cm)	Suppliers
Ag ink	0–5	10^5–10^6	DuPont, Elantas, Sun Chemical
CNT ink	10–50	10^2–10^3	Nanointegris, Brewer Science, Elantas
Stretchable Ag ink	20–100	10^4–10^4	Harima Chemicals, DuPont, nanoPaint
Stretchable CNT ink	20–100	10^2–10^3	Nanointegris, Brewer Science, nanoPaint
PEDOT: PSS	0–10	10^0–10^2	Heraeus, Agfa
Elastomeric conductive inks	20–200	10^0–10^2	Creative Materials, Sun Chemical
Graphene inks	10–50	10^2–10^3	Graphenea, Haydale, Graphenest
Liquid metal ink	100–300	10^4–10^6	LM ink, Liquid Wire Inc.

to improve adhesion, especially for non-polar surfaces. Adhesion strength should be maintained under bending, stretching, or other mechanical stresses to ensure durability in flexible applications.

- Functional inks must remain flexible to prevent cracking or loss of functionality under bending. This is especially important for wearable electronics or sensors that experience frequent bending. For applications involving stretchable electronics, the ink should be capable of stretching without losing electrical conductivity and adhesion. Functional inks should withstand repeated mechanical deformation without degradation, maintaining consistent performance over many bending or stretching cycles.
- Inks should maintain their properties over the expected range of operating environment (e.g., temperatures, humidity, UV light, solvents).
- Inks should be compatible with encapsulation materials that offer added chemical protection without reacting negatively or losing performance.
- The ink's viscosity and flow properties (rheology) should be optimized for the specific printing technique (e.g., screen printing, inkjet printing, gravure printing). Proper viscosity ensures uniform deposition, consistent pattern resolution, and adhesion to the substrate.
- Inks often require curing or sintering to develop full electrical conductivity or adhesion. Low-temperature curing inks are essential for flexible substrates that cannot withstand high temperatures, like PET or TPU.
- The ink should allow for precise patterning, enabling the creation of fine lines, high-resolution features, and accurate geometries in printed circuits or sensor elements.

- Inks used in wearables or biomedical devices should be biocompatible, non-toxic, and non-irritating to the skin, especially for devices in prolonged contact with the body. In biomedical sensors, inks should not react with biological tissues or fluids, ensuring stable performance and safe usage.

2.3 Fabrication Techniques

Fabrication techniques for PFS involve processes for printing functional materials onto flexible substrates to create electronic components and circuits [1]. The main fabrication sequence is depicted in Fig. 2.1, beginning with pre-printing, then printing, followed by post-printing and, finally, the integration with electronics. Not all these steps are to be followed, depending on the type of materials and techniques employed. These fabrication techniques offer flexibility, scalability, and versatility to produce PFS for various applications.

2.3.1 Pre-printing Techniques

Pre-printing techniques for PFS involve preparatory steps to optimize the substrate and materials before the actual printing process. These techniques are used for ensuring the adhesion and compatibility of the functional inks with the flexible substrate, leading to improved performance, reliability, and quality of the printed devices. These preparatory steps are essential for achieving consistent and reproducible results in PFS manufacturing processes. Some common pre-printing techniques are listed in Table 2.4.

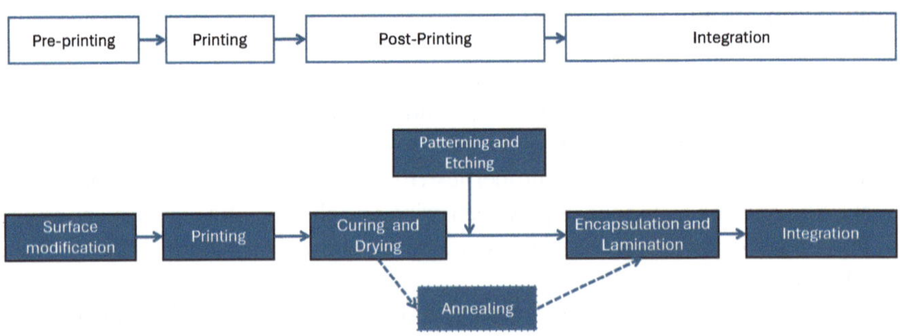

Fig. 2.1 Main fabrication steps for flexible printed electronics

Table 2.4 Main pre-printing techniques for flexible printed technologies

Pre-printing technique	Description	Methods and materials
Surface cleaning and preparation	Before printing, flexible substrates should be cleaned thoroughly to remove contaminants (e.g., dust, oils, residues) that may affect adhesion and print quality	Solvent wiping, plasma cleaning, ultrasonic cleaning, oxygen or atmospheric plasma, laser cleaning
Surface modification	Alter the surface properties of flexible substrates to improve adhesion, wettability, and compatibility with printed functional materials	Coatings, treatments (chemical, plasma), functionalization with adhesion promoters or coupling agents, roughening
Substrate coating	Involves applying thin functional coatings or primers (adhesion promoters, primer layers, barrier coatings) onto flexible substrates to enhance adhesion and compatibility with printed materials	Dip, spin, spray, knife coating, CVD, PVD
Substrate pretreatment	Modify the surface energy and chemistry of flexible substrates to promote ink adhesion and spreading; these methods improve the wetting and interaction between the substrate and the printed materials, leading to better print resolution and uniformity	Corona treatment, flame treatment, plasma treatment, etc.
Surface roughening	Create microstructures or texture patterns on flexible substrates to improve mechanical interlocking and adhesion of printed materials	Mechanical abrasion, laser ablation, etching processes
Preparation of printing inks and materials	Inks and functional materials should be prepared before printing to ensure compatibility with the substrate and desired printing process, including methods for ink thinning/thickening, particles dispersion, drying kinetics control	Ink formulations (rheology, surface tension, drying kinetics) optimized for specific printing techniques and substrates; mechanical or magnetic stirring; ultrasonication
Masking and patterning	Define precise regions or patterns on flexible substrates for subsequent printing steps	Masking materials (photoresists, tapes, stencils) used to selectively shield or expose substrate areas during printing or surface modification processes
Alignment and registration	Ensure accurate positioning and alignment of flexible substrates and printing masks or templates	Optical alignment systems, fiducial markers, mechanical fixtures; precise registration and alignment are relevant for multilayer printing processes

2.3.2 Printing Techniques

Printing technologies play a critical role in the fabrication of printed electronics, particularly in the field of PFS. These technologies enable the deposition of functional materials onto substrates in a precise and controlled manner, allowing for the rapid and cost-effective production of flexible, lightweight, and conformable devices. Printing techniques can be classified into different categories depending upon their principles of operation, deposition methods, and applications:

- physical contact or non-contact between the printhead and the substrate;
- printed pattern definition: by a mask/mold or digital;
- geometry to be printed: planar (2D) or non-planar (3D) techniques;
- special techniques.

Some of the key printing technologies used in printed electronics and sensors are shown in Table 2.5.

These printing technologies, along with advancements in functional materials and substrates, have led to significant progress in the field of printed electronics and sensors, enabling the development of innovative devices for a wide range of applications.

2.3.3 Post-printing Techniques

Post-printing techniques are crucial for enhancing the performance, durability, and functionality of PFS after the previous printing process. After the printing process, several post-printing steps may be necessary to further process and enhance the performance of printed flexible electronics. These steps depend on the specific application, materials used, and desired device characteristics. They involve various methods for curing and drying, annealing, encapsulation and lamination, and integration other electronic components to ensure that the printed devices meet the desired specifications and operational standards. Some of the main post-printing techniques used in printed flexible electronics are presented in Table 2.6.

Post-printing techniques are essential for transforming printed electronic components into fully functional, durable, and reliable devices. These techniques ensure that printed electronic devices can withstand various environmental conditions, maintain their performance over time, and be integrated seamlessly with other components.

Table 2.5 Main printing techniques for flexible printed technologies

Printing technique	Description	Main features
Planar 2D techniques		
Inkjet printing	Deposition of droplets of functional ink onto a substrate to create patterns or features; high resolution, scalability, print a wide range of materials; producing complex electronic circuits and sensors with fine features and high resolution	Non-contact and digital Ink viscosity: 1–10 cP Resolution: 20–30 μm Fine features, high-resolution
Screen printing	Versatile, widely used technique that involves forcing ink through a mesh stencil onto a substrate; printing thick layers of functional inks on various substrates; used for large-area applications, such as printed circuit boards, solar cells, and sensors	Contact, requiring a mask Ink viscosity: 500–5000 cP Resolution: 30–100 μm Large-area deposition Thick-film applications
Flexography	A flexible relief plate is used to transfer ink onto a substrate; used for high-speed, roll-to-roll printing of flexible electronics on continuous substrates; high throughput, suitable for large-scale production	Contact Ink viscosity: 50–500 cP Resolution: 50–200 μm Printing on flexible materials
Gravure printing or intaglio printing	Uses a cylindrical printing plate with recessed cells to transfer ink onto a substrate; capable of producing high-resolution images and fine features; precise patterns and intricate designs; used for security printing, flexible displays, and RFID tags	Contact Ink viscosity: 10–1000 cP Resolution: 10–50 μm High-speed, high-volume production of fine patterns
Offset printing	Widely used printing technique that involves transferring ink from a plate onto a rubber blanket and then onto a substrate; high throughput, suitable for mass production	Contact Ink viscosity: 100–1000 cP Resolution: 20–100 μm High throughput
Roll-to-roll (R2R) processing	Involves the continuous fabrication of flexible electronics on large rolls of flexible substrates; high-throughput production of flexible electronics, well-suited for scalable manufacturing	Flexible substrates High throughput In-line integration Scalable
Non-planar 3D techniques		
Pad printing	Versatile and widely used technique for transferring 2D images onto 3D objects	Contact printing Ink viscosity: 500–2500 cP Resolution: 100–150 μm Complex geometries

(continued)

Table 2.5 (continued)

Printing technique	Description	Main features
Aerosol jet printing	Non-contact method that uses an aerosolized stream of conductive ink particles to deposit material onto a substrate; precise, high-resolution printing of electronics on non-planar surfaces	Non-contact Ink viscosity: 1–1000 cP Resolution: 10–20 μm Complex geometries
Electrohydrodynamic jet printing (E-Jet printing)	Non-contact printing method that uses electric fields to control the deposition of functional materials onto a substrate; enables precise droplet formation and control, allowing for high-resolution printing	Non-contact Ink viscosity: 1–10 cP Resolution: <1 μm High precision and control
Laser direct writing	Non-contact, high-precision, high-resolution fabrication technique used to create patterns or structures on various materials by directly writing with a focused laser beam; localized material modification or ablation	Non-contact, but material ablation Resolution: <1 μm High precision, complex designs Microscale patterning
Water transfer printing or hydrographic printing or hydrodipping	Placing a water-soluble film, printed with the desired pattern, onto the surface of water in a dipping tank; the object to be decorated is then submerged into the water through the film, which wraps around and adheres to the object's surface; after drying, a protective coating is applied to enhance durability	Contact Suitable for 3D objects Versatile for complex patterns Applicable to a variety of materials and shapes
Other techniques		
Embossing, nanoimprint lithography	Replication of patterns onto flexible substrates using a mold or stamp; the mold is pressed onto the substrate, transferring the pattern through mechanical deformation or by imprinting functional materials	Contact High throughput Submicron resolution Patterning and imprinting on flexible materials
Soft lithography	Uses elastomeric stamps or molds (typically PDMS) to transfer patterns onto substrates; suitable for creating microstructures and printed electronics, e.g., microcontact printing, replica molding, microtransfer printing, solvent-assisted micromolding, etc.	Contact High resolution Microfabrication, patterning on a variety of substrates

2.4 Testing and Quality Control

Testing and quality control are essential post-printing steps to ensure the performance, reliability, and consistency of printed flexible electronics. Various characterization techniques, including electrical testing, optical microscopy, and surface profilometry, are used to evaluate device performance and identify defects or inconsistencies.

Table 2.6 Main post-printing techniques for flexible printed technologies

Post-printing technique	Description	Main features
Curing and drying techniques	Many printed materials require curing or drying to achieve their final properties. This step removes solvents, initiates chemical reactions, and enhances adhesion and mechanical strength **Curing methods** include thermal curing (heating in an oven), UV curing (exposure to ultraviolet light), chemical curing (reacting with a curing agent), and photonic sintering (pulsed light) In **Laser Sintering** focused laser beams are used to sinter conductive inks, providing precise control over the process and allowing for high-resolution patterning without damaging the substrate	**Thermal Curing**: Applying heat to cure inks, binding them to the substrate. This process enhances the mechanical stability and electrical conductivity of printed layers **UV Curing**: Using ultraviolet light to rapidly cure photoreactive inks and adhesives. This method is beneficial for materials that are sensitive to thermal curing **Photonic Sintering**: Intense pulsed light to sinter conductive inks at lower temperatures (suitable for temperature-sensitive substrates) **Laser sintering**: High precision and resolution, compatible with a wide range of materials
Annealing	**Annealing** is a heat treatment process used to improve the crystallinity, electrical conductivity, and mechanical properties of printed materials. Annealing is particularly important for organic semiconductors and conducting polymers to optimize charge transport and reduce defects	**Thermal Annealing**: Heating the printed electronic to enhance the performance of semiconductor and conductive inks **Rapid Thermal Processing (RTP)**: A high-temperature process applied for a short duration to improve the material properties without significantly affecting the substrate
Encapsulation and lamination	**Encapsulation** involves covering the printed electronics with protective layers to shield them from environmental factors such as humidity, oxidation, and damage **Lamination** techniques, such as thermal bonding or adhesive lamination, are used to encapsulate printed components between flexible films or coatings	**Slot-Die Coating**: continuous deposition of inks onto flexible substrates through a narrow slot-shaped die. Coating large-area substrates with thin, uniform layers **Spin coating**: a small amount of liquid coating is applied on the center of the substrate that is then rotated at very high speeds, spreading the coating material by centrifugal force **Conformal Coating**: apply a thin conformal coating (e.g., silicone, acrylic) to protect the printed patterns (e.g., spray, dispensing, dip coating) **Dry-film Lamination**: pressure or heat is used to bind a thin protective adhesive film to the printed substrate (by roll or knife pressure)

(continued)

Table 2.6 (continued)

Post-printing technique	Description	Main features
Patterning and etching	Used to define precise features, traces, or electrodes on printed materials; **laser ablation, chemical etching**, and reactive ion etching (RIE) are used to selectively remove material layers and create desired patterns	**Laser Ablation and Laser Scribing**: high-energy laser beams to selectively remove or pattern materials through vaporization or melting, leaving behind patterned features. Laser scribing involves the localized heating and ablation of thin films to create conductive traces and patterns
Cutting and trimming	**Laser Cutting** uses laser beams to cut or trim printed electronics to the desired shape and size with high precision **Die Cutting** uses a special die to cut shapes out of flexible materials, by pressing it into the material **Other techniques**: mechanical cutting (knife cutting), water jet cutting, ultrasonic cutting	**Laser cutting**: Non-contact, high precision, suitable for intricate designs and small features, fast processing, suitable for high-throughput production, can cut a wide range of materials **Die cutting**: Suitable for mass production, consistency producing uniform shapes and sizes, cost-effective, good for simple and repeatable shapes
Stretching and forming	**Stretching** techniques to create stretchable printed electronics by printing on pre-stretched substrates and then relaxing them **In Forming**, printed electronics are shaped to fit specific applications, such as conforming to the contours of a wearable device or integrating into complex 3D structures (e.g., in-mold electronics)	**Mechanical Stretching**: uniform stretching, controlled force **Thermoforming**: Heat-activated stretching, Mold-based shaping allowing complex shapes with high precision **Elastomeric Substrate Stretching**: High elasticity, printed or embedded electronics can stretch with the substrate **Vacuum/Pressure Forming**: Intricate and detailed shapes, ensuring the substrate conforms tightly to the mold; pressure produces higher details and accuracy

Functional testing consists in verifying the performance of printed electronics by testing their electrical, mechanical, and thermal properties. Automated testing setups can be used for large-scale production, thus ensuring that integrated sensors provide accurate and reliable measurements. Printed patterns may include additional pads for local assessment of the quality of printed patterns (Fig. 2.2). These can be used in in-line testing, performing automated testing during the production process to verify sensor functionality and performance.

Also, post-production calibration procedures can be implemented, adjusting sensor outputs to match known reference standards, ensuring accuracy and consistency.

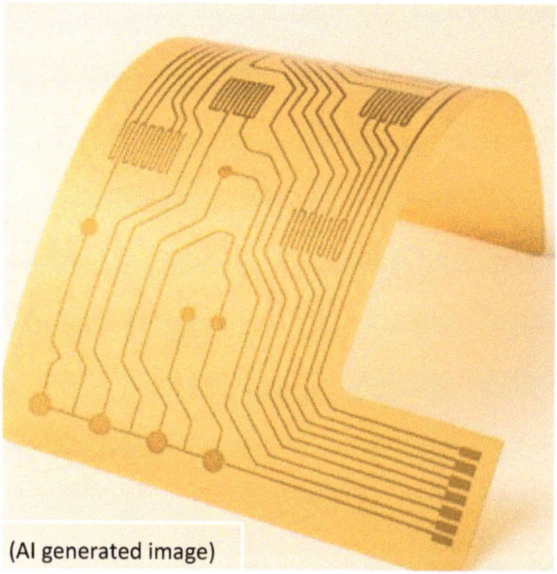

Fig. 2.2 Printed pads for testing and quality control

2.5 In-Line and Roll-To-Roll Production

Roll-to-roll (R2R) technology is noted for its potential in industrial-scale production of flexible electronics (Fig. 2.3). It is a highly efficient and scalable method for producing PFS and other electronic devices. This continuous processing technique involves the fabrication of electronic components on a flexible substrate that is unwound from a roll, processed through various stages, and then rewound onto another roll.

Main steps of R2R processing are (Fig. 2.3):

1. **Unwinding**: The flexible substrate is unwound from the roll and fed into the processing line.
2. **Surface Treatment**: Techniques like corona, plasma treatment, or chemical cleaning are applied to improve ink adhesion and surface properties.
3. **Printing**: Various printing methods can be used to deposit functional inks onto the substrate in precise patterns to create the sensor elements and circuitry, such as gravure printing, flexographic printing, screen printing, inkjet printing.
4. **Drying and Curing**: Printed inks are dried and cured using thermal, UV, or laser processes to solidify the materials and enhance their electrical properties.
5. **Integration**: Additional components, e.g., chips, batteries, can be integrated onto the flexible substrate using methods like pick-and-place and adhesive bonding.
6. **Encapsulation and Lamination**: Protective layers are added to shield the sensors from environmental damage.
7. **Inspection and Testing**: Automated systems check the printed sensors for defects and test their functionality to ensure they meet performance standards.

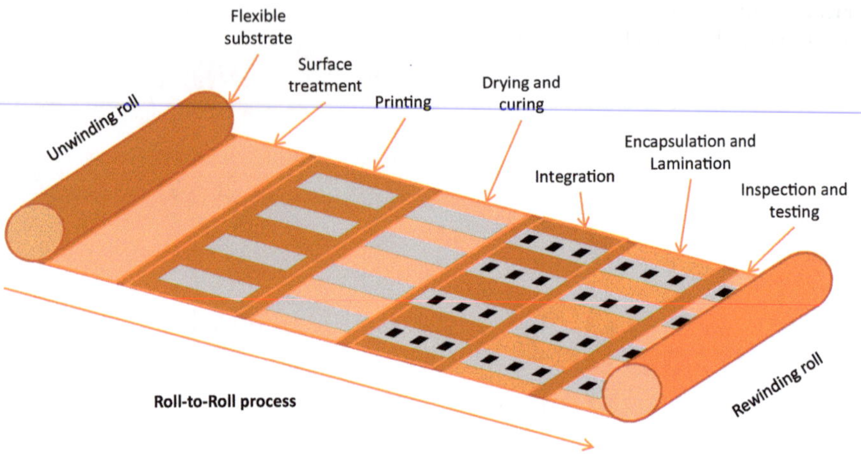

Fig. 2.3 R2R processing and possible sequential task/process stations

8. **Rewinding**: Processed substrate is rewound onto a roll for easy handling, storage, and transport.

R2R offers several advantages for PFS, such as:

- **Scalability**: High throughput and continuous production make it ideal for large-scale manufacturing.
- **Cost-Effectiveness**: Reduced material waste and lower energy consumption compared to traditional batch processing.
- **Customization**: Easy adaptation to different designs and functionalities by changing the printed patterns and materials.
- **Environmental Benefits**: Potential for using eco-friendly materials and processes, contributing to sustainability goals.

2.6 Additive Manufacturing

Additive manufacturing techniques, such as 3D printing, can be used to fabricate 3D structures and devices directly onto flexible substrates. Materials such as conductive filaments, functional inks, and flexible polymers can be used to create complex, multimaterial structures for flexible electronics applications, including FPS.

3D printing enables the fabrication of 3D objects, layer-by-layer from a digital model, and it has also been applied to the fabrication of electronic devices and sensors [7], enabling the fabrication of complex geometries, integrating multiple materials, and providing high customization [8]. These techniques allow for the precise placement of conductive, semiconductive, and dielectric materials, essential for creating high-performance flexible sensors, and it can be used to print conductive traces, sensors, and other electronic components directly onto 3D objects.

3D printing techniques allow the manufacturing of customized sensor designs tailored to specific applications and requirements. They can produce complex shapes and structures that are difficult or impossible to achieve with traditional manufacturing methods. They also allow for the integration of multiple materials within a single device, enhancing functionality and performance. Various techniques such as Fused Deposition Modeling (FDM), Stereolithography (SLA), Selective Laser Sintering (SLS), and Direct Ink Writing (DIW) are used, as shown in Table 2.7.

Table 2.7 3D printing techniques for printed flexible sensors

3D printing technique	Description	Main features
Fused deposition modelling, FDM	Involves extruding thermoplastics through a heated nozzle to build objects layer-by-layer Materials: thermoplastic elastomers (TPE), TPU, conductive filaments	Material versatility, cost-effective, widely accessible, good for creating flexible and durable sensors
Stereolithography, SLA	Uses a UV laser to cure liquid photopolymer resins layer-by-layer Materials: photopolymer resins, flexible resins, conductive resins	High resolution, smooth-surface finish, detailed sensor structures
Digital light processing (DLP)	Similar to SLA, DLP uses a digital light projector to cure liquid resin. It cures entire layers at once, rather than point-by-point, like in SLA Materials: photopolymer resins, flexible resins, conductive resins	Faster layer curing compared to SLA, produces high-detail and smooth-surface components
Selective laser sintering, SLS	Uses a laser to fuse powdered materials layer-by-layer Materials: Nylon, TPU powders	Works with a variety of powdered materials, good mechanical properties, no need for support structure
Direct ink writing, DIW	Extrudes ink or paste materials through a nozzle to create three-dimensional structures, layer-by-layer Materials: conductive inks, silicone-based, hydrogels	Materials versatility, intricate designs, embedding conductive pathways

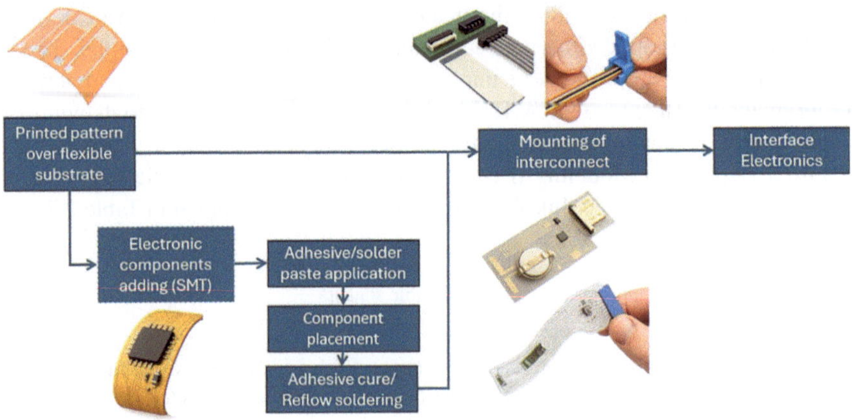

Fig. 2.4 Connecting the FPS with interface electronics

2.7 Integration of Printed Flexible Sensors with Electronics

PFS may require integration with electronic components such as microcontrollers, communication modules, battery management system, etc., which, typically, are rigid components (Fig. 2.4).

PFS can be directly linked to the interface electronics with a connector. Depending upon the design, the printed pads can be crimped or bonded with a FCP/FPP connector (Flexible Flat Cable/Flexible Printed Circuit) onto the flexible substrate. Integration may also require the bonding of electronic components in the flexible substrate. The main integration steps are: (a) connectors, (b) mounting and bonding of electronic components, and (c) packaging to assemble functional electronic systems with printed components.

2.7.1 Connectors for Printed Flexible Sensors

Connectors for PFS are crucial for ensuring reliable electrical connections while keeping the flexibility and durability of the sensors. There are various types of connectors for printed films (Table 2.8), typically based on Surface Mount Technology (SMT), i.e., soldering/bonding the connector directly onto the surface of the substrate. Printed patterns and connectors shall be designed together and take into consideration flexibility, stretchability, and durability. By selecting appropriate connectors and using reliable connection techniques, PFS can function effectively in various applications, maintaining their performance and durability.

Table 2.8 Types of connectors for PFS

Type of connector	Description	Advantages	Example
Pin and socket connectors	Traditional connectors with pins and sockets adapted for flexible circuits	Offer secure connections and can be designed for repeated mating cycles	
Flexible flat cables, FFC Flexible printed circuits, FPC	FFC are flat, flexible ribbon-like cables, very thin and flexible, with multiple conductive tracks FPC connect flexible thin substrates	Highly flexible, thin, compact; can handle large number of connections in a compact form	Figure 2.5
Crimp flex connector	Uses a crimping tool to attach connectors to the flexible sensor leads	Requires specific crimping tools and connectors designed for flexible circuits	Figure 2.5
Zero insertion force, ZIF, connectors	Use a lever or slider to insert and lock the flexible cable in place without applying insertion force	Provide secure and reliable connections with minimal mechanical stress on the flexible substrate	
Pogo pins	Spring-loaded pins that provide temporary or permanent connections between flexible sensors and other circuits	Allow for easy connection and disconnection, suitable for testing and prototyping	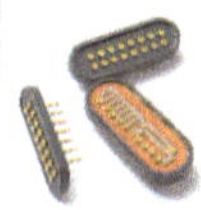

FPC and FFC connectors are specially designed for flexible electronic applications (Fig. 2.5). They show excellent flexibility, a low-profile design minimizing space, a high density of connections in a minimal footprint, and ease of installation.

Conformal connectors enable the connection of electronic components and circuits on non-planar, flexible, and 3D surfaces, being an essential technology for flexible electronics. These connectors conform to the shape of the substrate and maintain electrical connectivity even when bent, twisted, or stretched, and include:

- **Flexible connectors** make use of flexible connectors or conductive tapes to connect different parts of flexible circuits.

Fig. 2.5 Main types of connectors

- **Stretchable connectors** can elongate and return to their original shape without losing electrical continuity, such as BIND (biphasic, nano-dispersed) interface, and stretch-safe (a magnetic connector).
- **Stretchable and bendable connectors** that maintain electrical connections even when the device is flexed or stretched.

Stretchable and bendable connectors are flexible electronic components designed to maintain connectivity even when stretched, bent, or twisted. They can be based on:

- **Conductive polymers and elastomers**, like silicone embedded with conductive particles (e.g., Ag, C), offer both stretchability and electrical conductivity; these materials are molded into connector shapes or printed as flexible circuits.
- **Liquid metal conductors**, like gallium-based alloys are often used for stretchable circuits because they can move and shift while retaining electrical conductivity; encapsulated in a flexible substrate, these alloys form highly durable stretchable connectors.
- **Textile-based connectors** are textile fibers that are woven or coated with conductive materials to create fabric-like connectors
- **Printed stretchable circuits** use inkjet printing of conductive ink on stretchable substrates for producing flexible connectors.

Printing flexible connections by additive manufacturing techniques has been proposed (by NanoDimension Technologies Ltd. Company, https://www.nano-di.com/) by 3D printing a series of multilayered rigid PCB, connected through, also printed, flexible conductive connectors [7]. This eliminates the need to connect the circuits, saves space, and provides connections with high mechanical performance.

2.7.2 Mount Technologies of Electronic Components

The integration of additional components can be done by conventional SMT that attaches traditional electronic components onto printed flexible circuits, combining the best of both printed and conventional electronics. Pick-and-place equipment allows the precise placement of components, adding discrete components such as resistors, capacitors, chips, LED, and other electronic components, to the printed flexible circuits. Specialized machines adapted to handle and place components on flexible substrates without causing damage are used.

2.7.3 Bonding Techniques

Bonding techniques are required to attach the electronic components securely to the printed flexible substrate. This can be done by resorting to solders or electrically conductive adhesives. These connections should maintain electrical connectivity while allowing for some mechanical flexibility. To ensure reliable integration of these components onto flexible substrates, various bonding techniques are employed, as described in Table 2.9.

Table 2.9 Techniques for bonding components to PFS

Technique	Description	Considerations
Direct soldering	Soldering flexible printed sensor pads directly to the connection points on a flexible circuit	Requires careful control of soldering temperature to avoid flexible substrate damages
Reflow soldering	Adapted reflow ovens with controlled temperature profiles to prevent warping or damage to flexible materials. Reflow ovens can be altered with advanced temperature control systems, including multiple heating zones and rapid cooling capabilities, and real-time monitoring of temperature profiles	Precise control of temperature throughout the reflow process to ensure that the flexible substrates are not exposed to excessive heat, which can cause warping, shrinkage, or degradation
Low-temperature soldering	Use of low-melting-point solder alloys (120–180 °C) to minimize thermal stress	Typically based on Tin–Bismuth (SnBi), Tin–Bismuth-Silver (SnBiAg), Tin-Indium (SnIn), or Indium Alloys (e.g., InAg)
Laser soldering	Focused laser beams for precise and localized soldering, reducing thermal impact on surrounding areas. The non-contact nature of the process eliminates mechanical stress on the substrate, making it ideal for delicate and flexible materials	The laser beam can be precisely controlled to target specific solder joints, providing localized heating without affecting the entire substrate
Heat and pressure bonding	Applying heat and pressure to bond conductive materials to flexible substrates, ensuring strong and reliable connections	Suitable for mass production but requires precise control to ensure reliable connections
Conductive adhesives	Used to bond interconnects to substrates and components; contain conductive particles, providing both mechanical bonding and electrical connectivity	Thermal adhesion uses heat to bond the placed components to the substrate; use of UV light to cure and harden the conductive adhesive

(continued)

Table 2.9 (continued)

Technique	Description	Considerations
Anisotropic conductive films, ACF	Adhesive films embedded with conductive particles that create electrical connections only in the z-axis. Conduct electricity in one direction and are used for connecting components to flexible circuits	Provide reliable electrical connections while maintaining mechanical flexibility and are suitable for fine-pitch applications
Zebra bonding	Creates electrical connections using Zebra connectors, which are made from alternating conductive and non-conductive layers, often silicone elastomer; used to connect fine-pitch components in electronic assemblies	Suitable for applications that require bending and compression; they can handle very fine pitch, making them ideal for connecting components with small contact areas
Ultrasonic welding	Using high-frequency ultrasonic vibrations to bond conductive materials	Provides strong and reliable connections, especially for thermoplastic substrates

2.7.4 *Hybrid Electronics*

As mentioned above, PFS can be integrated with conventional electronics (e.g., rigid PCB technology) by developing hybrid systems that seamlessly combine PFS with traditional electronic components. Flexible hybrid electronics (FHE) integrates PFS with rigid silicon-based components, providing a balance between flexibility and advanced functionality. FHE extends the design flexibility of conventional rigid electronics. It enhances the performance of printed and flexible components by integrating high-performance traditional electronic components and allows the seamless integration into objects. It is also cost effective due to the potentially low manufacturing costs of the printing techniques.

Chapter 3
Design Considerations

Designing PFS requires careful consideration of materials, sensor configuration, printing technologies (including pre- and post-printing operations), environmental factors, and others to ensure that the sensors perform reliably while maintaining flexibility. A balance must be made between performance, durability, and flexibility while ensuring reliable function under environmental and mechanical stress. Understanding sensing mechanisms under different stimuli is important for carefully choosing materials, optimizing geometry, and enabling efficient data acquisition from the sensor.

3.1 Sensing Mechanisms

PFS uses a variety of sensing mechanisms to detect and measure different types of physical, chemical, and biological stimuli. The sensing mechanism affects the design of the PFS. The primary types of sensing mechanisms are (Fig. 3.1):

Resistive Sensing: Changes in electrical resistance due to mechanical deformation (e.g., bending, stretching) or changes in the environmental conditions (e.g., temperature, humidity) are measured to infer the sensed parameter.

Magnetoresistive Sensing: Detect changes in electrical resistance in the presence of a magnetic field. These sensors are useful for detecting magnetic fields, and for position sensing.

Piezoresistive Sensing: Relies on changes in the electrical resistance when the sensor is deformed. Applied stress or pressure alters the electrical resistance of the material, which can be measured to determine the force or pressure applied. These sensors are commonly used in tactile sensing and pressure monitoring.

Capacitive Sensing: Variations in capacitance, resulting from changes in the distance between conductive layers or in the dielectric properties of a sensor material

J. C. Viana, *Primer on Printed Flexible Sensors*,
https://doi.org/10.1007/978-3-032-06554-4_3

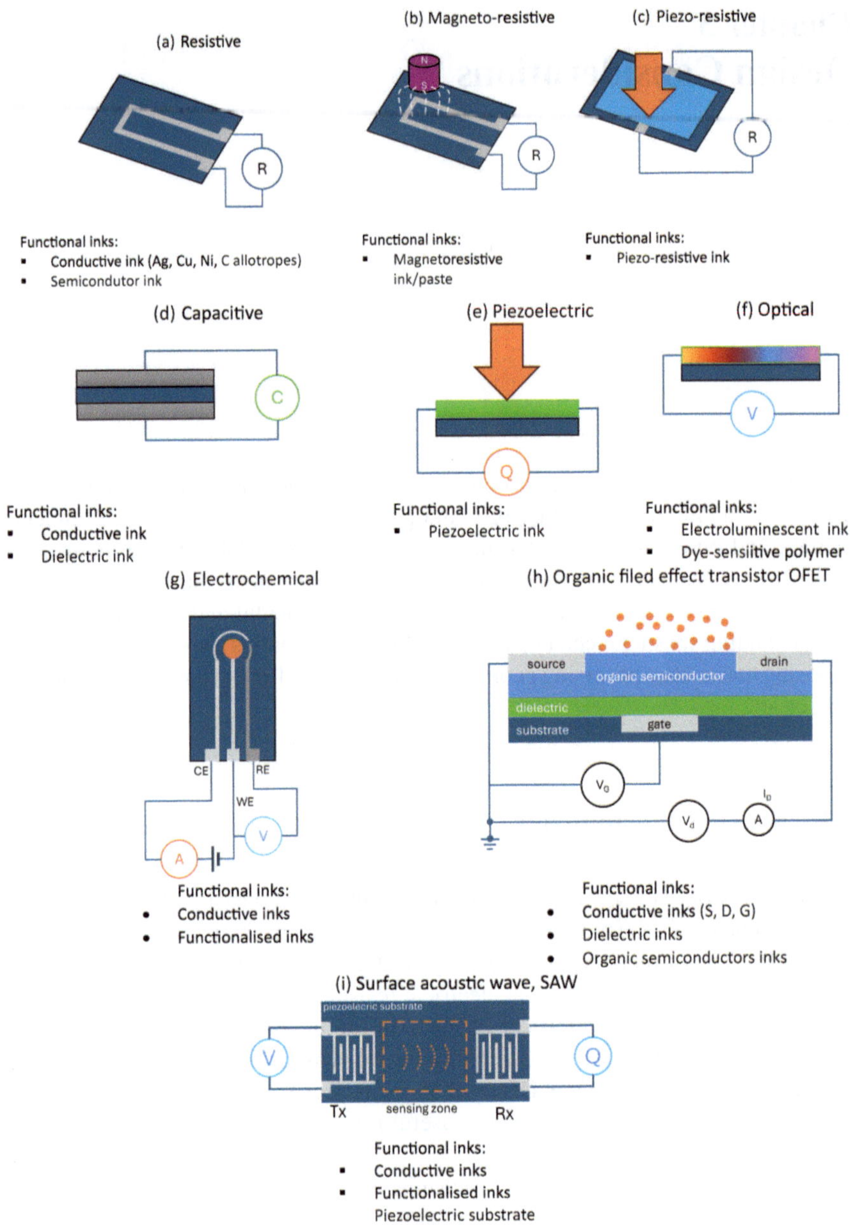

Fig. 3.1 Main types of sensing mechanisms for PFS

in response to external stimuli, are detected and measured. Capacitive sensors are widely used for touch screens, proximity sensors, and humidity detection.

Piezoelectric Sensing: This mechanism is based on the generation of an electric charge in response to mechanical stress. Piezoelectric materials produce a voltage when subjected to deformation, making them suitable for applications in vibration sensing, energy harvesting, and dynamic pressure sensing.

Optical Sensing: Changes in optical properties (e.g., absorbance, fluorescence, reflectance) of sensing materials in response to specific analytes or environmental conditions are detected. Optical sensors detect changes in light properties such as intensity, wavelength, or polarization. They often incorporate materials that change their optical properties in response to external stimuli, e.g.: (a) photodetectors measure light intensity changes; (b) fluorescence detects light emitted by fluorescent tags or dyes; and (c) absorbance measures the amount of light absorbed by the material.

Electrochemical Sensing: Changes in the electrical properties such as current, voltage, or impedance in response to chemical reactions. Electrochemical sensors detect analytes through redox reactions at the electrode surface and convert the chemical information into an electrical signal. These sensors typically use electrodes printed with conductive inks and functionalized with specific sensing materials. The quantities assessed can be: (a) amperometric, if measuring current produced by redox reactions; (b) potentiometric, when voltage varies due to ion concentration differences; or (c) impedimetric, if changes in electrical impedance are measured.

Organic Field-Effect Transistor, OFET: this sensing mechanism relies on changes in the electrical properties of an organic semiconductor layer in response to target molecules or environmental conditions. When a target analyte (such as a gas, ion, or biomolecule) interacts with the OFET's active layer or dielectric, it alters the charge carrier mobility or threshold voltage. This change affects the drain current, enabling the OFET to transduce the interaction into a measurable electrical signal. This mechanism is highly sensitive to surface interactions, making OFET ideal for applications in chemical, biological, and environmental sensing.

Surface Acoustic Waves (SAW) sensing: uses acoustic waves traveling along the surface of a piezoelectric material to detect physical, chemical, or biological changes in the environment. When an input signal (Tx) is applied to an IDE on a piezoelectric substrate, it generates surface acoustic waves that propagate along the surface and are influenced by changes in the surface properties or the surrounding medium. When the waves reach a second IDE (Rx), any variations in the wave's amplitude, phase, or velocity, caused by interactions within the sensing zone, are detected and converted into an electrical signal. SAW sensors are known for their high sensitivity, rapid response, and ability to operate in harsh environments, making them suitable for applications in gas sensing, biosensing, pressure, and temperature monitoring.

Each of these sensing mechanisms can be implemented using various materials and fabrication techniques tailored to specific applications, leveraging the advantages of PFS technologies.

3.2 Design Parameters of Printed Flexible Sensors

The main design parameters of PFS are crucial for ensuring optimal performance, flexibility, adaptability, and durability. The integration of sensor's form factor (configuration and dimensions) and flexibility (materials and thickness) can be made based on several design strategies.

3.2.1 *Form Factor*

The form factor of printed sensors refers to their physical dimensions, shape, and configuration, which are crucial aspects in their design and application. Since printed sensors can be tailored to fit various applications, their form factor is often dictated by the specific needs of the device/system they are to be integrated with. The form factor of printed sensors is a blend of their physical design, intended application, and integration requirements. Printed sensors can be designed in virtually any shape to conform to the surfaces they will be attached to or integrated within. This includes complex geometries that traditional rigid sensors cannot accommodate, e.g.,

Thin and Lightweight: One of the primary advantages of PFS is their thinness, typically just a few micrometers to a fraction of a millimeter thick, making them ideal for applications where space is limited or where the sensor must be unobtrusive.

Flexible Strips and Tapes: These are elongated, narrow sensors that can be wrapped around objects or limbs, useful in applications requiring conformity to cylindrical shapes.

Patches: Often used in medical and fitness applications, patches are flat, typically with a skin-adhesive layer that allows them to attach directly to the body for monitoring physiological parameters.

Textile Integration: Sensors printed directly onto fabrics, enabling the creation of smart garments that can sense body movements, temperature, or other environmental factors.

Large-Area Mats and Skins: Used in smart building applications or as skin for robots, these sensors cover large areas and can detect pressure distributions, impacts, etc.

Customized 3D Shapes: With advancements in 3D printing and flexible materials, sensors can now be integrated into 3D structures tailored for specific applications, such as components that fit within irregular mechanical assemblies.

PFS comes in various designs and shapes tailored to specific applications and sensing mechanisms. Figure 3.2 shows some sensor configurations for different sensing mechanisms.

Resistive Layouts

The simpler design of a resistive sensor consists of two electrodes separated by a sensing material whose resistance changes with a physical stimulus. The sensitive

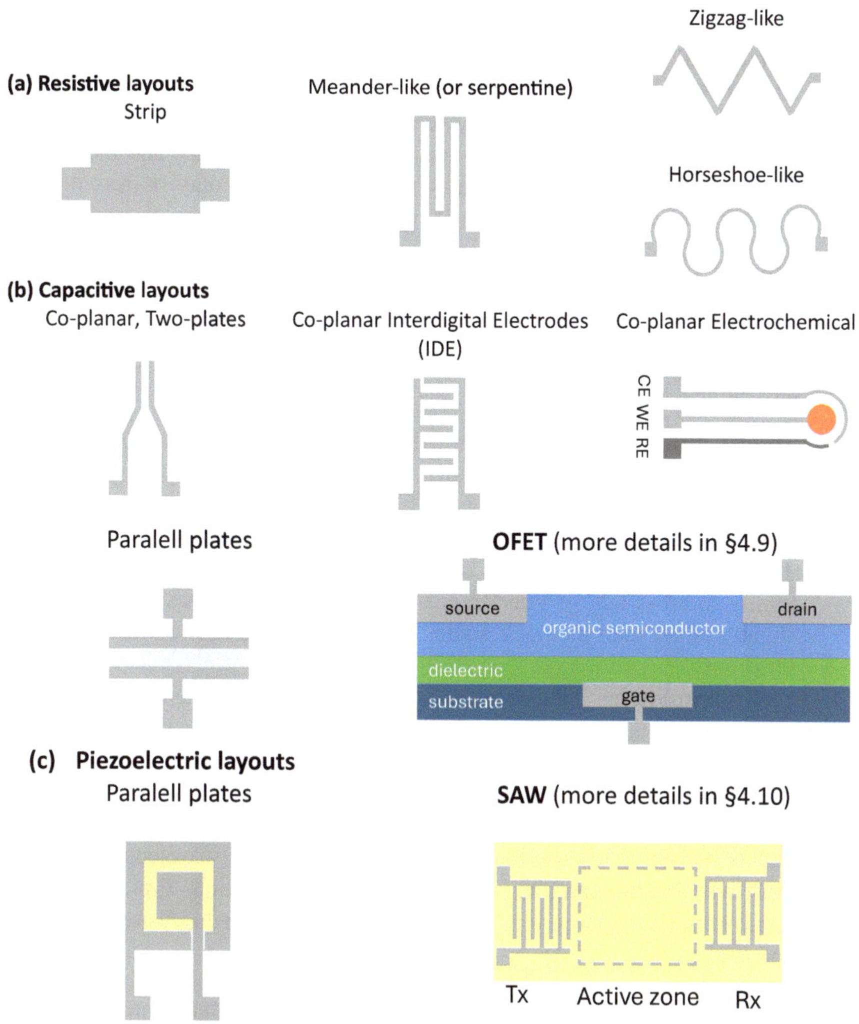

Fig. 3.2 Typical sensors configurations (WE—working electrode, CE—counterelectrode, RE—reference electrode, Tx—transmitter IDE, Rx—Receiver IDE; in yellow a piezoelectrical material)

material can have different shapes: a simple strip, a meander-like pattern, zigzag-like, or horseshoe-like patterns.

- **Strip resistive patterns** offer simplicity and effectiveness in detecting changes in resistance due to physical or environmental changes. Main geometrical design parameters are:
  - **Strip Width** (W) affects electrical resistance and current-carrying capacity. Narrower strips increase resistance and sensitivity but may be more prone to

breaks or inconsistencies. Typical width range is from 10 to 500 μm, depending on the application and fabrication capabilities.
  - Longer **strip length** (L) increases the electrical resistance and sensitivity. The typical length range is 1 mm to several centimeters, depending on the design.
  - **Strip Thickness** (t) affects the overall electrical resistance and mechanical flexibility. Thicker strips can carry more current, but may be less flexible. Typical thickness range from of 100 nm to several mm.

- A **meander-like (or serpentine) pattern** increases the surface area and the length of the resistive element. This allows more interaction and exposure to the physical stimulus, resulting in improved sensitivity, more accurate detection, and quicker and more pronounced responses. Main geometrical design parameters of serpentine tracks are:

  - **Track width** (W) affects the electrical resistance and current-carrying capacity. Narrower tracks increase resistance and sensitivity but may be more prone to breaks or inconsistencies. Typical widths vary between 10 and 500 μm, depending on the application and fabrication capabilities.
  - **Track Spacing** (S) affects the overall footprint of the sensor and its susceptibility to cross-talk or interference. Typical spacing values equal to or are slightly larger than the track width, depending on the desired resolution and sensitivity.
  - Larger **amplitudes** (height of the meander pattern from the centerline to the peak, A) can increase the overall length of the conductive path and enhance sensitivity to strain. Typical amplitude range is from 100 μm to several millimeters, depending on the design and application.
  - **Wavelength** (distance between successive peaks of the meander pattern, λ) affects the sensor's sensitivity and resolution. Shorter wavelengths increase sensitivity but may require higher precision in fabrication. Typical distances range from 100 μm to several millimeters, depending on the desired sensitivity and application.
  - **Overall length** (L) **and width** (W_total) determine the overall electrical resistance and the sensor's ability to cover the target area. Typical dimensions are variable, based on application requirements.

The meander pattern can accommodate some bending, stretching, and twisting without breaking or losing functionality. It allows for a compact design that maximizes the functional length of the sensor within a limited area. This design can help mitigate signal noise by averaging out variations over the extended length, resulting in more stable and reliable readings. In applications involving temperature monitoring or thermal responses, the meander pattern can help to dissipate heat more effectively, preventing localized overheating and ensuring consistent performance.

- **Zigzag patterns** are often used to enhance the sensor's flexibility, to increase the effective length of the conductive path, and to improve the sensor's ability to detect changes in electrical resistance due to physical deformation or environmental changes. Main geometrical design parameters of zigzag patterns are:

- **Track width** (W) affects the electrical resistance and current-carrying capacity. Narrower tracks increase resistance and sensitivity but may be more prone to breaks or inconsistencies. Typical widths range from 10 to 500 μm, depending on the application and fabrication capabilities.
- **Track spacing** (distance between adjacent tracks, S) affects the overall footprint of the sensor and its susceptibility to cross-talk or interference. Typical spacing values equal to or are slightly larger than the track width, depending on the desired resolution and sensitivity.
- Larger **amplitudes** (height of the zigzag pattern from the centerline to the peak, A) can increase the overall length of the conductive path and enhance sensitivity to strain. Typical amplitude values range from 100 μm to several millimeters, depending on the design and application.
- **Wavelength** (distance between successive peaks of the zigzag pattern, λ) affects the sensor's sensitivity and resolution. Shorter wavelengths increase sensitivity but may require higher precision in fabrication. Typical distances range between 100 μm and several mm, depending on the desired sensitivity and application.
- **Overall length** (L) and **width** (W_total) determine the overall resistance and the sensor's ability to cover the target area. Typical dimensions are variable, based on application requirements.

- **Resistive horseshoe patterns** show enhanced flexibility and stretchability. They allow the sensor to bend and stretch more easily compared to straight-line patterns. This makes them suitable for applications that require significant deformation or conformability to curved surfaces. The design distributes strain more evenly across the sensor, reducing the likelihood of failure or degradation under mechanical stress. Main geometrical design parameters of horseshow pattern are:

  - Larger **Horseshoe Radii** (R) can reduce stress concentration and increase flexibility but may decrease the overall electrical resistance and sensitivity. Typical radii vary from 0.5 mm to several millimeters, depending on the application and fabrication capabilities.
  - **Track Width** (W)—the same as the other patterns mentioned above.
  - **Track Spacing** (S)—identical to the other patterns mentioned above.
  - **Number of Horseshoe Turns** (N) is the number of repeated horseshoe shapes in the pattern. More turns increase the effective length of the conductive path, enhancing sensitivity and the electrical resistance. Typical number of horseshoes turns range from 1 to several turns, depending on the design and application.

The horseshoe shape can endure repeated bending and stretching cycles without significant performance degradation, increasing the sensor's longevity. By distributing stress, the horseshoe pattern helps prevent the formation of cracks, which can occur in more rigid designs. The curved segments of the horseshoe pattern can concentrate strain in specific areas, enhancing the sensor's sensitivity to mechanical changes. The pattern can be designed to have a higher density of sensing elements in a given area, improving the resolution and accuracy of measurements.

Capacitive Layouts

Capacitive sensors consist of a pair of parallel electrodes with a dielectric material in between. In a co-planar design, interdigitated or adjacent conductive tracks lay on the same plane.

- A **co-planar** single **two-plates capacitor** is not a common design for a printed capacitor sensor (IDE configuration being more usual). Common layouts include spiral, grid, and interdigitated patterns. This design is simplified, but still requires careful consideration of design parameters to achieve high sensitivity, flexibility, and stability:
  - **Electrode width** (W) and **spacing** (S) between adjacent electrodes control the effective capacitance. Narrower gaps and wider electrodes increase capacitance, improving the sensor's sensitivity, but may reduce range.
  - **Longer electrode** traces provide a larger interaction area, which enhances sensitivity and overall capacitance.
  - **Dielectric material**: In co-planar sensor design, the dielectric constant of the substrate material influences the electric field generated by the sensor, impacting detection range and sensitivity.
- The structure of an **InterDigitated Electrode (IDE)** comprises two comb-like electrode arrays interlaced with each other. This design creates a uniform electric field between the fingers, leading to a more consistent and reliable sensor response. It enhances the efficiency of the sensing mechanisms. Designing of IDE for PFS involves several key geometrical parameters that influence the sensor's performance, and sensitivity. Main design parameters of IDE are:
  - **Electrode Width** (W): affects the capacitance value and sensitivity. Narrower electrodes can increase the sensitivity by increasing the electric field strength but may be more challenging to fabricate precisely. Typical widths vary from 10 to 100 μm, depending on the application and fabrication capabilities.
  - **Electrode Spacing** (S): smaller gaps can increase sensitivity by creating a stronger electric field but can also increase the risk of short circuits. Typical spacing values range from 10 to 100 μm, similar to W.
  - **Electrode Length** (L): Longer electrodes provide a larger sensing area, enhancing sensitivity and signal strength. Typical lengths vary from 1 mm to several millimeters, depending on the substrate size and application requirements.
  - **Number of Fingers** (N): A higher number of fingers increases the overall surface area and interaction sites, improving sensitivity and signal strength. Typical N values vary widely, from a few pairs to several hundred pairs, depending on the application.
  - **Dielectric Material and its Thickness**: The dielectric constant (ε) and thickness (t) of the dielectric material affect the capacitance and sensitivity. A higher

dielectric constant increases capacitance, but might introduce nonlinearity, while thickness influences the distance of the electric field penetration. The thickness varies from few micrometers to hundreds of micrometers, depending on the application.

- An **electrochemical sensor** detects analytes through redox reactions at the electrode surface. It comprises working (WE), reference (RE), and counterelectrodes (CE) printed on a flexible substrate. Larger surface areas of the WE enhance sensitivity by providing more active sites for redox reactions. Nanostructuring or modifying its surface can increase the effective surface area. The distance between WE, RE, and CE affects the efficiency of the electrochemical cell and the diffusion of ions. Proper spacing is crucial for minimizing electrical resistance and maximizing sensitivity. The overall design and dimensions of the sensor affect the sensor's integration into devices, its mechanical properties, and the sample volume requirements. Key design parameters and considerations for electrochemical sensors are:
  - **Electrode Configuration**, i.e., the arrangement and design of the working, reference, and counterelectrodes, determine the efficiency of electron transfer, the sensitivity, and the overall performance of the sensor. Common configurations are planar electrodes and IDE.
  - **Electrode Surface Area** is the effective surface area of WE. Larger surface areas enhance sensitivity by providing more active sites for redox reactions. Nanostructuring or modifying the electrode surface can increase its effective surface area.
  - **Electrode Spacing**, the distance between the working, reference, and counterelectrodes, affects the efficiency of the electrochemical cell and the diffusion of ions. Proper spacing is crucial for minimizing resistance and maximizing sensitivity.
  - **Electrode Materials** (conductive material used for WE, RE, and CE) affect the electrical conductivity, the stability, and the specificity of the sensor. The choice of material can also influence the reaction kinetics and electrode surface properties. Common materials for RE are: Au, Pt, C (graphite, graphene, CNT), Ag/AgCl.
  - **Electrode Modification** corresponds to the functionalization of the electrode surface with specific materials to enhance selectivity and sensitivity and to improve the specificity for target analytes. It can also enhance electron transfer rates. Common electrode modifications are: enzyme immobilization, metal nanoparticles, conductive polymers, molecularly imprinted polymers (MIP).
  - **Electrolyte Composition**, i.e., the ionic medium that facilitates the redox reactions affects electrical conductivity, stability, and reaction kinetics. The pH and ionic strength of the electrolyte can also impact sensor performance. Commonly used electrolytes are: phosphate-buffered saline (PBS), potassium chloride (KCl), and sulfuric acid (H_2SO_4).
  - **Membrane or Separator** is a selective barrier that separates WE from the sample. It provides selectivity by allowing specific ions or molecules to

reach the electrode surface while blocking others. It can also protect the electrode from fouling. Common materials are: ionomers (e.g., sulfonated tetrafluoroethylene-based fluoropolymer-copolymer or Nafion®), cellulose acetate, polyvinyl alcohol (PVA).

- **OFET sensors** operate as a capacitor, with a conducting channel between the source and the drain (S and D) electrodes. The applied voltage at the gate controls the amount of charge carriers (electron or holes) flowing through the system. The OFET adopts typically the architecture of a thin-film transistor (TFT), with S and D electrodes deposited onto a thin layer of an organic semiconductor, and a thin-film dielectric material deposited between the semiconductor and the gate electrode. Due to its specific configurations, design considerations of OFET are presented in Sect. 4.9.4.

Piezoelectric Layouts

- A **piezoelectric sensor** typically involves a layer of piezoelectric material (e.g., PVdF, PZT, ZnO) sandwiched between electrodes. Key design parameters of piezoelectric sensors are:
  - **Thickness of the piezoelectric layer** (t) affects the sensor's resonant frequency, sensitivity, and mechanical flexibility. The thickness varies from micrometers to millimeters, depending on material type and the application.
  - The **overall size and shape** of the sensor affect the resonant frequency, sensitivity, and compatibility with the application. Typical sizes are around few millimeters, depending on the application
  - **Surface Acoustic Wave (SAW)** sensors use acoustic waves traveling along the surface of a piezoelectric material to detect changes in physical, chemical, or biological conditions. There are several configurations of a SAW sensor, and due to their specificity, their design considerations are presented in §4.10.3.

3.2.2 Flexibility and Stretchability

Flexibility refers to the sensor's ability to bend, twist, and stretch, often without losing functionality. This characteristic is essential for applications involving movement or variable shapes. Bendability is crucial for any application where the sensor needs to conform to curved surfaces. The minimum bend radius must be defined to ensure the sensor does not crack or lose connectivity, losing its functionality. The bending radius depends upon Young's modulus and ductility of the substrate, and its thickness, but also upon the flexibility of the functional inks and their adhesion to the substrate. For applications requiring significant deformation, the sensor must withstand stretching without loss of functionality. Stretchability is essential for wearable applications where the sensor must accommodate skin deformation or muscle expansion. Materials and designs (e.g., using zigzag or horseshoe-like tracks) are chosen to allow for improved elasticity. Durability under continuous flexion stresses is also relevant, mainly for dynamic sensor applications. The sensor must withstand repeated bending or stretching cycles without degrading in performance, which is particularly challenging for conductive materials.

3.2.3 Integration of Form Factor and Flexibility

The integration of form factors and flexibility of PFS involves several design strategies:

a. **Material Selection**: using substrates like PI, PET, or TPU that offer both flexibility and durability. Flexible conductive inks may include Ag, C, or graphene-based compounds formulated to maintain electrical conductivity under strain.
b. **Structural Design**: Incorporating patterns that enhance flexibility, such as zigzag or horseshoes-like conductive tracks, which allow the material to stretch and bend without breaking the conductive paths.
c. **Layer Configuration**: Optimizing the thickness and arrangement of the different layers in the sensor to balance flexibility with other necessary properties, such as encapsulation for environmental protection or adhesion layers for mounting.
d. **Integration with Other Components**: Ensuring that connectors and interfaces also support the flexibility of the entire system, which may involve flexible printed circuit boards or specialized connectors that tolerate movement.

The form factor and flexibility of printed sensors are designed to meet specific application needs, allowing these sensors to be integrated seamlessly into diverse environments and devices. The challenge lies in maintaining functionality and reliability while maximizing flexibility and conforming to desired product shapes.

3.2.4 Other Considerations

PFS can be fabricated over large areas as needed, enabling applications such as wall-sized touch panels or smart skin for robots. In applications like pressure mapping, the density of sensing elements within a given area is also important. Printed sensors allow for the customization of sensor density based on specific needs, controlling the spatial resolution. Integration capability and stackability are other relevant design considerations. Sensors can be designed to be stackable with other electronic components in a multilayer configuration, allowing for the integration of multiple functionalities within the same footprint. The form factor must also consider how the sensor will connect with other electronic components (interface with electronics), such as connectors and control units, ensuring that connections are both physically and electrically compatible. Sensors often require encapsulation to protect against environmental factors like humidity, chemicals, and mechanical abrasion. The form factor must accommodate any additional layers or coatings needed for protection, which can change the flexibility or stretchability of the printed sensor. Especially in consumer and medical applications, the appearance and tactile feel of the sensor can be important. Portability and disposability of the sensor are also to be considered. For portable devices, the form factor should be compact and lightweight. In applications like health monitoring patches, the sensor might be designed for single use and therefore must be cost-effective and disposable, with raised eco-sustainability issues.

3.3 Design of Interconnects

Interconnects provide electrical connections between various components of an electronic system, such as integrated circuits (IC), electronic components, and other electronic devices. They enable the transmission of electrical signals and power throughout the system, ensuring proper communication and functionality. In printed electronics, the interconnects are the conductive tracks connecting different electronic components, the vias (typically plated holes that allow electrical connections between different layers of a substrate), and pads (that are exposed areas where components are soldered/bonded). In the case of flexible printed electronics, they must be able to withstand mechanical stresses such as bending, stretching, and twisting. The design of the interconnect patterns must optimize the flexibility of the substrate and of the functional ink. Some best practices for the design of interconnectors for FPS are shown in Table 3.1.

3.4 Design Approach for PFS

Designing PFS involves a multifaceted approach that considers both the functionality, the constraints imposed by the materials' flexibility and the printing technologies. The design and development of PFS involve several steps, as depicted in Fig. 3.3.

1. **Sensor Requirements**
    - Purpose and Functionality: Clearly define what the sensor is intended to measure.
    - Performance Metrics: Specify the sensitivity, range, resolution, and accuracy required.
    - Operating Conditions: Define the environmental conditions under which the sensor must operate, including temperature ranges, environmental exposure (e.g., water, UV, chemicals).

Table 3.1 Guidelines for the design of interconnectors for PFS

Design best practice	Description	Example
Mechanical reinforcement	Reinforce the connection points on flexible sensors with additional layers or stiffeners to prevent damage during connection and use	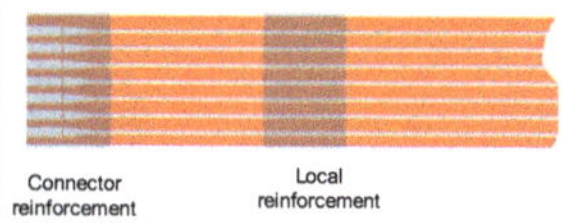
Stress relief	Design the layout to include stress relief features such as loops or bends near the connectors to minimize mechanical stress on the connection points; hexagonal hatched zones can better distributed stresses	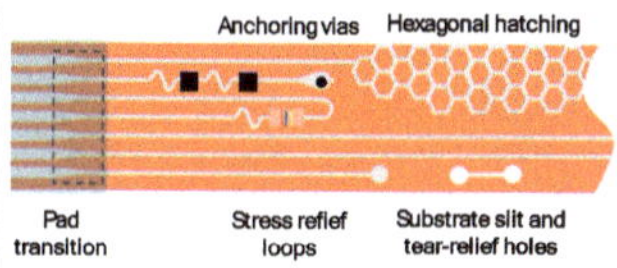
Meander, zigzag or horseshoes-like patterns	Designing interconnects in meander, zig-zag, or horseshoes-like patterns allows them to be stretches, to reduce local mechanical stresses and to flex without breaking	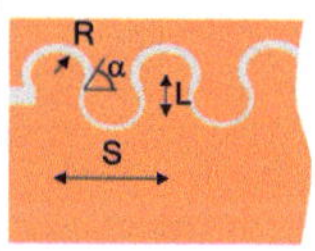
Island-bridge structures	Components are placed on rigid "islands" connected by flexible "bridges" of conductive material, allowing for movement and flexibility, and stability of the components	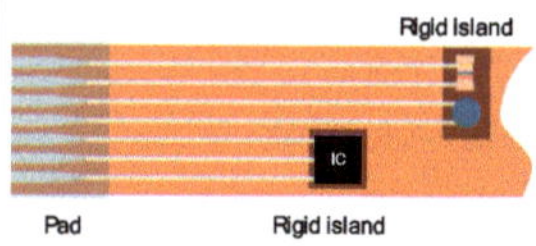
Kirigami and origami techniques	These techniques involve cutting and folding materials to create interconnects that can stretch and conform to various shapes	
Testing and validation	Perform rigorous testing, including mechanical stress and environmental tests, to ensure that the connectors maintain reliable performance under all expected conditions	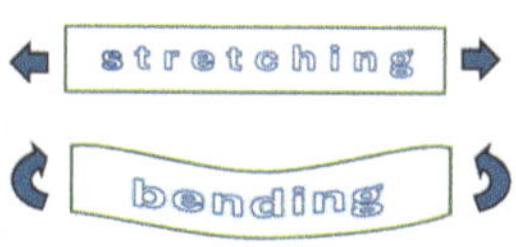
Prototyping and iteration	Prototype different connector designs and configurations to find the optimal solution for specific applications and requirements	

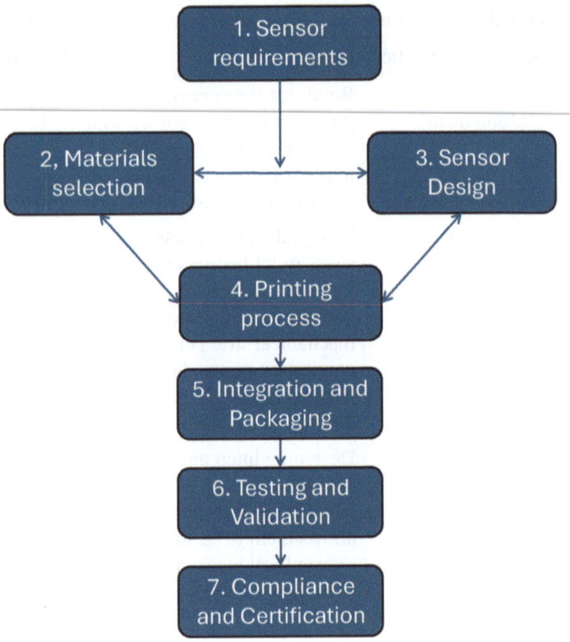

Fig. 3.3 Design and development of PFS

2. **Materials Selection**
   - Substrate: Choose a flexible substrate that suits the application environment and desired flexibility (e.g., PET, PI, or TPU). The substrate should also be compatible with the chosen printing technology (including pre- and post-printing operations).
   - Functional Inks: Select conductive, dielectric, and other functional inks based on the electrical requirements and chemical stability. Inks must maintain their properties when stretched or bent.

3. **Sensor Design**
   - Circuit Layout: Design the circuit layout to maximize the performance and minimize disruptions caused by bending or stretching; this often involves strategic placement of conductive paths to avoid high-stress areas.
   - Interconnects and Junctions: Design robust interconnects that can tolerate flexural loadings. Use serpentine, zigzag, wavy, or horseshoes-like patterns for conductive tracks to increase their stretchability.
   - Scaling and Dimensions: Optimize the thickness and width of the printed conductive tracks to balance electrical performance and mechanical flexibility.
   - Sustainable Design: Use eco-friendly materials (e.g., biodegradable, renewable, or non-toxic materials) that can mitigate sustainability issues. However, finding materials that offer the necessary performance characteristics while being environmentally friendly remains a challenge.

4. **Printing Process**

   - Printing Technique Selection: Choose a suitable printing technique (e.g., screen printing, inkjet printing, aerosol jet printing) based on the product shape (planar, 3D), required resolution, material compatibility, production volume, and cost.
   - Layer Management: Plan the sequence and integration of different material layers, considering drying times and interaction between layers.
   - Prototyping: Prototype the sensor to test the design and materials in real conditions, adjusting the design as needed based on performance feedback.

5. **Integration and Packaging**

   - Electronic Integration: Integrate the PFS with necessary electronics, ensuring that connections are flexible and secure. Consider using flexible PCB or directly printed electronics.
   - Encapsulation: Protect the sensor from environmental factors and mechanical damage with suitable encapsulation materials that do not compromise flexibility.

6. **Testing and Validation**

   - Mechanical Testing: Perform bending, stretching, and fatigue tests to ensure the sensor can withstand physical stress without failure.
   - Functional Testing: Test the sensor under controlled conditions to validate its accuracy, sensitivity, and response time against the defined requirements.
   - Iterative Development: Use testing results to refine the sensor design, materials, and manufacturing process iteratively.

7. **Compliance and Certification**

   - Standards Compliance: Ensure the sensor meets any relevant standards, particularly if used in regulated industries like healthcare or automotive.
   - Certification: Obtain necessary certifications for safety, quality, and environmental impact.

By methodically addressing these areas, designers can develop PFSs that are not only effective in their intended applications but also robust and reliable under their operational environments. By combining the right **materials**, **design**, and **fabrication techniques**, optimized printed flexible sensors for maximum sensitivity, durability, and cost-efficiency can be produced. Table 3.2 shows typical materials and printing techniques for some applications.

Table 3.2 Materials (substrates and inks) and printing techniques used in some applications

Applications	Substrates	Functional inks	Printing techniques
Healthcare and wearable sensors	**PDMS, Ecoflex**, or **TPU** for skin compatibility and flexibility	**Graphene, PEDOT: PSS**, or **AgNP** for biocompatibility and conductivity	**Inkjet** or **aerosol jet printing** for high precision in biosensor design
Environmental monitoring sensors	**PET** or **PI** for stable performance in variable temperatures	**Metal Oxides** (ZnO, TiO_2) for gas detection; **Ag inks** for stable conductivity	**Screen printing** or **flexographic printing** for large-scale production
Smart textiles and flexible electronics	**TPU, Ecoflex**, or **Textile-integrated layers** for durability	**Carbon-based inks, graphene**, or **CNT** for enhanced flexibility	**Screen printing** or **3D Printing** for integrating circuits directly into fabric
Industrial sensors (pressure, force, and gas sensors)	**PI** or **PET** for heat resistance and durability	**Graphene, CNT**, or **Cu inks** for stable conductivity	**Aerosol jet printing** or **3D printing** for complex pressure sensor geometries

3.5 Sustainability Issues

PFS offers numerous advantages in terms of functionality and application, but there are several sustainability issues associated with their development, use, and disposal to be considered. Table 3.3 resumes some of these sustainability issues applicable to PFS.

The development of PFS must address the applicable environmental regulations. Compliance with standards ensuring that PFS meets environmental regulations and standards is essential. This includes compliance with restrictions on hazardous substances (e.g., RoHS, REACH) and adherence to guidelines for waste management. Furthermore, conducting comprehensive lifecycle assessments (LCA) to understand the environmental impact of PFS from raw materials extraction to product end-of-life disposal can help identify areas for improvements and guide the development of more sustainable practices [9]. Addressing these sustainability issues requires a holistic approach, including the development of new materials and manufacturing techniques, the improvement of product design, better recycling and disposal methods, and the adherence to environmental regulations. Most of the abovementioned sustainability issues of PFS are common to other types of sensors. Table 3.4 compares PFS and conventional sensors in terms of main sustainability issues.

The environmental benefits of PFS are many. Generally, they allow for: (a) reduced material usage and waste generation; (b) lower energy consumption in production; (c) potential for using environmentally friendly materials and processes. Nevertheless, the lifecycle impact needs careful assessment, particularly for newer materials, and potential trade-offs in performance and durability compared to conventional electronics.

Table 3.3 Main sustainability issues for flexible printed sensors

Sustainability issue	Impact	Description
Materials	Toxicity and environmental impact Resource scarcity	Materials used in PFS, such as certain polymers, metals, and nanomaterials (e.g., Ag, CNT) can be toxic and pose environmental hazards during manufacturing, usage, and disposal. Others are derived from scarce or non-renewable resources, contributing to resource depletion and supply chain vulnerabilities
Manufacturing	Energy consumption Waste generation and management	The production of PFS can be energy-intensive, especially if high-temperature processes or specialized equipment are required Manufacturing processes often generate waste materials (e.g., solvents, chemicals), which need to be managed and disposed of properly to minimize environmental impact
Lifecycle and durability	Product longevity Degradation and performance	PFS may have shorter lifespans compared to traditional sensors, leading to more frequent replacement and increased material consumption Environmental factors such as humidity, temperature, and mechanical stress can degrade sensor performance over time
End-of-life management	Recycling and disposal e-waste	PFS can be challenging to recycle due to the complex mix of materials used. Improper disposal can have environmental impact The rapid growth of flexible electronics contributes to the overall increase in electronic waste (e-waste), posing significant challenges for waste management and recycling infrastructure

Table 3.4 Comparison of sustainability issues for conventional and printed flexible sensors

Sustainability issue	Conventional sensors	Printed (flexible) sensors
Materials use	Bulkier, non-renewable materials such as silicon, metals, ceramics, and plastics Often require rare earth elements, which have significant environmental and ethical issues	Organic and inorganic materials deposited in thin layers, reducing material usage and weight Can be made from biodegradable or recyclable materials, potentially reducing environmental impact
Production processes and energy consumption	Involves subtractive manufacturing processes, where materials are etched, leading to higher waste Require multiple, complex steps, increasing the resource and time intensity of production Production processes are energy-intensive, involving high-temperature treatments Manufacturing clean rooms and other infrastructure contribute to high energy demands	Utilize additive manufacturing techniques that deposit materials only where needed, reducing waste Produced on flexible substrates, enabling R2R manufacturing which is more efficient and scalable Involve lower energy for manufacturing due to simpler production processes Production can occur at lower temperatures, further reducing energy consumption

(continued)

Table 3.4 (continued)

Sustainability issue	Conventional sensors	Printed (flexible) sensors
Waste generation	Significant waste generated from etching, cleaning, etc. High volume of e-waste at end-of-life, with many components not being easily recyclable	Lower waste generation during production (additive processes) Potential for reduced electronic waste if sensors are designed to be biodegradable or recyclable
End-of-life disposal	Often difficult to recycle due to the complex assembly and the use of non-recyclable materials E-waste is a significant environmental problem (hazardous materials pose risks to health and environment)	Possibility of designing for easier recycling or biodegradability Flexibility and lightweight nature may facilitate improved end-of-life management

Chapter 4
Types of Printed Flexible Sensors

A PFS is designed to detect and measure various physical, chemical, or biological parameters. These sensors are manufactured using printing techniques on flexible substrates using functional inks. Simply, a PFS consists of a flexible substrate (polymer, elastomer, textile, paper), printed electrodes and pads in an electrically conductive ink, and a printed sensing material (Fig. 4.1).

PFS are developed with unique sensing capabilities suited to different applications, being able to detect and measure various parameters/stimulus, such as strain, force and pressure, proximity, temperature, humidity, optical properties, gases and chemicals, electrochemical responses, biological parameters. This can be done by adopting several sensing mechanisms that measure distinct responses to the applied stimuli.

PFS are lightweight, bendable, stretchable, and conformable to various surfaces and shapes. Their flexibility and adaptability make them ideal for a wide range of physical environment sensing and monitoring, enabling automated responses and valuable data collection applications.

4.1 Strain Sensors

Printed flexible strain sensors are flexible devices that measure strain in various structures. The flexibility of these sensors allows them to conform to different surfaces and bend without losing functionality. They often use changes in electrical resistance, capacitance, or optical properties to quantify the strain. In the following text, only electrical strain sensors will be discussed, being optical sensors dealt with in Sect. 4.6.

J. C. Viana, *Primer on Printed Flexible Sensors*,
https://doi.org/10.1007/978-3-032-06554-4_4

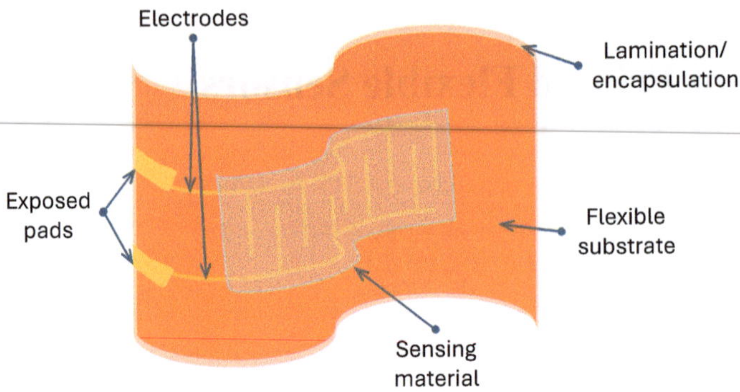

Fig. 4.1 Typical components of a PFS

4.1.1 Piezoresistive Strain Sensors

A piezoresistive strain sensor responds to a change in the electrical resistance, R, of a conductive/resistive ink when deformed. The relationship between the electrical resistance and the strain, ε, is dependent upon the type of sensing material. Several sensing mechanisms have been explored (Fig. 4.2).

- **Metal conductors, geometrical effect (Fig. 3.2a)**

For an isotropic metal conductor of electrical resistivity ρ, length l, and cross-section A, the deformation causes a variation of R of:

$$\frac{dR}{R} = \frac{d\rho}{\rho} + \frac{dl}{l} - \frac{dA}{A} = \frac{d\rho}{\rho} + \frac{dl}{l}(1 - 2\vartheta). \tag{4.1}$$

The first term ($d\rho/\rho$) corresponds to the contribution of the variations upon the intrinsic piezoresistivity of the material, whereas the last term ($dl/l(1 - 2\nu)$) is related to the contribution of the variations of sample dimensions during deformation (geometrical effect), involving the Poisson coefficient, ν.

The gauge factor, GF, is an important figure of merit of a strain sensor, representing the sensitivity of the sensor's electrical output to the input mechanical strain. It quantifies how much the electrical property of the sensor changes per unit strain. For a resistive strain sensor, GF is typically defined as the ratio of the relative change in resistance ($\Delta R/R_0$) to the mechanical strain, ε:

$$GF = \frac{\Delta R/R_0}{\varepsilon} = \frac{\Delta\rho/\rho_0}{\varepsilon} + (1 + 2\vartheta) \tag{4.2}$$

where the indices 0 refer to the initial zero strain variables. Higher GF corresponds to a greater sensitivity. Ideally, GF shall remain constant over the range of strains the sensor is designed to measure. The GF of a metallic strain sensor is around 2.

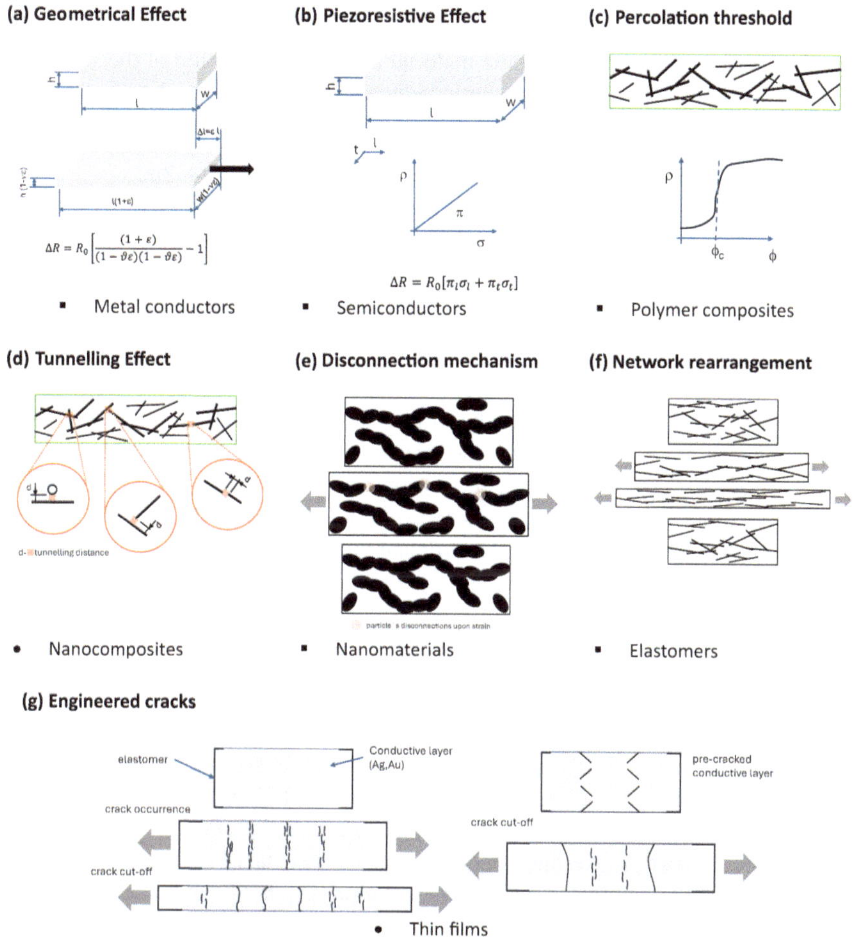

Fig. 4.2 Main strain sensing mechanisms

- **Semiconductors, piezoresistive effect (Fig. 3.2b)**

For a semiconductor, the change in electrical resistance is attributed to alterations in its band structure, affecting the movement of charge carriers (electrons and holes) within the material. The number of charge carriers is highly sensitive to strain, giving a high piezoresistive effect to semiconductors. When a semiconductor material is strained, several changes take place:

I. lattice deformation occurs affecting the band structure of the material;
II. the carrier mobility is affected, and scattering mechanisms can change charge mobility significantly; and also,
III. the density of charge carriers is altered.

This results in a high piezoresistive coefficient that varies significantly in different directions, depending on the specific material and the construction of the sensor.

Considering Eq. (4.2), *GF* of a semiconductor strain sensor can be given by:

$$GF = \frac{\Delta\rho/\rho_0}{\varepsilon} + (1 + 2\vartheta) = \frac{(\pi_L + 2\pi_T)\sigma}{\varepsilon} + (1 + 2\vartheta)$$
$$= (\pi_L + 2\pi_T)E + (1 + 2\vartheta) \quad (4.3)$$

where π_L and π_T are the piezoresistive coefficients in the longitudinal and transverse directions of the semiconductor, σ is the applied stress, and E is Young's modulus of the semiconductor material. The *GF* for semiconductors can vary depending on factors such as the type of semiconductor material used, its crystal structure, its doping levels, and the orientation of the strain relative to the crystal lattice. Generally, semiconductor materials with high piezoresistive coefficients tend to have higher *GF*. The *GF* of semiconductor strain sensors are notably high, typically ranging from 100 to 300, with electrical resistances of about 200 Ω–2 kΩ. Despite the high GF, semiconductors sensors show a significant nonlinearity and are also sensitive to temperature.

- **Conductive polymer composites and nanocomposites**

(a) **Percolation mechanism (Fig. 3.2c)**

For conductive polymer composites, consisting of a dielectric polymeric matrix with incorporated electrically conductive particles, the electric response is dependent upon the concentration of particles, according to the percolation theory. A percolation threshold is achieved when, upon increasing the filler concentration, a conductive path is established and a drastic reduction in the electrical resistance occurs. The electrical conductivity of the composite near the percolation threshold can be described by a power law:

$$\sigma = \sigma_0(X - X_C)^B \quad (4.4)$$

where X is the filler volume fraction, X_C the volume fraction of the filler at the percolation threshold, and B is a critical exponent. This latter is dependent on the properties of the filler (e.g., dimensionality and geometry). The sensitivity of the electrical resistance variation with strain is higher at the percolation threshold. Generally, this sensitivity increases with the reduction of the amount of conductive particles. These strain sensors can achieve high *GF* values, by inducing microstructural changes that affect the conductive pathways.

(b) **Tunneling effect (Fig. 3.2d)**

In piezoresistive strain sensors based on nanomaterials (like CNT, quantum dots, or nanowires) deformation can influence the tunneling of charge carriers through the material. Tunneling effect is significant when the filler particles are close but not in direct contact. This happens near or slightly below the percolation threshold, where the filler particles form a network with small gaps between them. As the distance between nanoparticles changes due to strain, the probability of charge carrier tunneling alters, leading to changes in the electrical conductivity. The tunneling resistance, R_{tunnel}, depends upon the height and width of the tunnel barrier (e.g., matrix properties), and it may be defined as [10]:

$$R_{\text{tunnel}} = \frac{h^2 d}{A q_e^2 \sqrt{2m\tau}} e^{\frac{4\pi d}{h}\sqrt{2m\tau\pi}} \quad (4.5)$$

where h is Planck's constant, d is the distance between conductive particles (tunneling distance), q_e is charge of an electron and m its mass, A is a pre-exponential factor, and τ represents the tunnel barrier height. Considering Eq. (4.4), the electrical conductivity of conductive nanocomposites near the percolation threshold can be described by:

$$\sigma = \sigma_0 (X - X_C)^B + \sigma_t \quad (4.6)$$

where σ_t is the electrical conductivity due to the tunneling effect. If the filler particles are close enough, the tunneling probability increases, leading to a measurable electrical conductivity even slightly below X_C. The tunneling current density through a barrier can be estimated using the Simmons formula (for rectangular potential barriers). The tunneling conductivity can be related to J by:

$$\sigma_t = \frac{Jd}{V} \propto d \cdot \exp\left(-\frac{2d\sqrt{2m\varnothing}}{h}\right) \quad (4.7)$$

where d is the barrier thickness, V is the applied voltage, m is the electron mass, ϕ is the barrier height (in eV), $\hbar$ is the reduced Planck's constant. Equation 4.7 shows that the tunneling conductivity decreases exponentially with increasing barrier thickness and barrier height. In disordered nanocomposites, the conductivity due to tunneling between nanoparticles is often described by:

$$\sigma_t = \sigma_0 \cdot \exp(-\beta d) \quad (4.8)$$

where σ_0 is a pre-exponential factor, β is the tunneling decay constant (typically, of $\beta \approx 1 - 2\ \text{nm}^{-1}$), and d is the average interparticle separation.

(c) **Disconnection mechanism (Fig. 3.2e)**

Disconnection is another resistive sensing mechanism for flexible strain sensors, primarily for those based on nanomaterials (e.g., polymer nanocomposites). In some sensing materials, such as Ag nanowires, ZnO nanowires, CNT, or graphene, conductive particles can slide against each other under the applied strain. The disconnection (due to sliding) decreases the overlapped area of the sensing particles and creates less conductive pathways, increasing the overall electrical resistance. This disconnection occurs because of weak interfacial bonding between the conductive nanoparticles and the significant elastic mismatch between nanoparticle and the polymeric matrix. The electromechanical response of the sensor is determined by these disconnection-reconnection processes. If the contacting conductive particles continually slide until a small gap is generated between them, tunneling effects can occur.

- **Elastomers, network rearrangements (Fig. 3.2f)**

Elastomers can be filled with conductive particles to become more electrical conductors. Their electrical resistance varies with the amount of incorporated conductive particles according to the percolation theory. However, due to their hyperelastic behavior, elastomers allow large deformations elastically recovered. This leads to the rearrangement of the conductive particles within the elastomer, changing the electrical pathways and contact resistance between them, and affecting the electrical resistance response with strain. This means that different mechanisms can be present during deformation, depending upon the type and concentration of conductive particles, the type of elastomeric matrix, the fabrication process and the strain level.

- **Thin films, engineered cracks (Fig. 3.2g)**

Engineered cracks can significantly enhance the sensitivity of strain and pressure sensors. The mechanism is bioinspired by spiders' sensory system on their slit sensilla organ. Cracks are intentionally created in a conductive thin film deposited on a stretchable substrate. Upon stretching microcracks are separated or generated in the conductive film. The number of conductive pathways decreases and the electrical resistance largely increases with strain (evidencing a very high GF). By adjusting the density, distribution, shape, size, and orientation of engineered cracks, the electrical conductivity of this sensing film can be precisely controlled. Also, the percolation threshold can be fine-tuned. Several techniques can be used to create engineered cracks in conductive polymer composites, such as (a) mechanical methods (cutting, stretching); (b) thermal methods (thermal cycling, laser ablation); and (c) chemical methods. GF up to 2000 at low strains are obtained. But cracks do not self-heal after strain.

(d) **Multiple mechanism**

Flexible resistive strain sensors can operate based on multiple mechanisms simultaneously. This multiple mechanism approach can enhance the strain sensitivity of the sensor, can provide a wider range of detectable strain, and can improve the accuracy and the reliability of the sensors.

4.1.2 Piezoelectric Strain Sensors

A piezoelectric strain sensor uses a pooled piezoelectric ink or piezoelectric strip printed or bonded over/to a dielectric substrate to generate an electric charge Q ($Q = d_{ij}F_i$, where d_{ij} is the piezoelectric coefficient in the polarization axis direction,

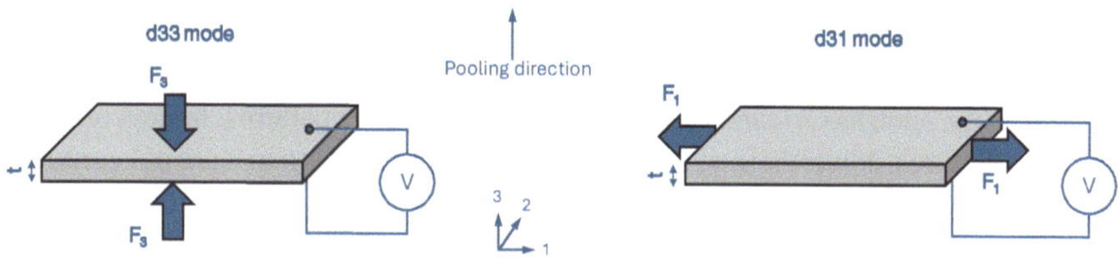

Fig. 4.3 Piezoelectric charge (strain) coefficients d_{33} and d_{31} of a printed piezoelectric ink

i, when the force is applied in direction *j*) in response to an applied dynamical load (F_j).

Printed piezoelectric inks must be pooled before being used, i.e., subjected to a very high electric field that aligns all the dipole moments in the same direction of the electrical field (this being carried out at temperatures below the Curie temperature). This is called polarization direction (typically direction 3). A piezoelectric material can operate in different modes, the most used being shown in Fig. 4.3.

- In d_{33} (longitudinal) operating mode, the force is along axis 3, parallel with the polarization axis (3); i.e., the piezoelectric material is loaded on the same surface from which the electrical charges are collected (the voltage is measured in the same direction of the applied force).
- In d_{31} (transverse) operating mode, the charge is collected along the polarization direction (3), but the force is applied at right angles (axis 1) to this direction

Typically, $d_{33} > d_{31}$, following the empirical relation: $d_{33} \approx -2.5d_{31}$ (mainly for PZT materials). Following Fig. 4.14, the relationship between strain (in direction 3, ϵ_3) and voltage generated (V_c) by the piezoelectric sensor is given by:

$$V_c = \frac{td_{33}E\epsilon_3}{\varepsilon_{33}} \tag{4.9}$$

where ε_{33} is the in-plane dielectric permittivity, t is the thickness of the piezoelectric material, d_{33} is piezoelectric coefficient in operating mode 33, and E is Young's modulus of the piezoelectric material. Other polarization and loading directions will generate different V_c values. The GF in the longitudinal mode, GF_{33}, is given by:

$$GF_{33} = \frac{d_{33}t}{\varepsilon_0\varepsilon_{33}\epsilon_3 A} \tag{4.10}$$

where ε_0 is the permittivity of free space, ε_{33} is the permittivity of the piezoelectric material, A and t are its area and thickness, respectively.

Piezoelectric strain sensors produce an instantaneous voltage when strained, only measuring dynamic strains (a static strain does not generate voltage). They also do not require continuing power to operate, reducing powering needs. Nevertheless, piezoelectric materials with high piezoelectric coefficients are ceramic-based, being rigid and brittle (e.g., PZT). Furthermore, piezoelectric polymers (e.g., PVdF) are flexible, but not so stretchable, requiring structural design solutions for improved stretchability, such as Origami or Kirigami designs, zigzag patterns, honeycomb and helical structures. Nevertheless, flexible piezoelectric strain sensors are not common.

4.1.3 Capacitive Strain Sensors

A capacitive strain sensor responds to a change in capacitance, C, with deformation. In these capacitive-based printed strain sensors, deformation causes a change in the distance or overlap between conductive electrodes separated by a dielectric material.

$$C = \varepsilon_0 \varepsilon_r \frac{lw}{d} \tag{4.11}$$

where l, w, and d are the length, width, and thickness of the dielectric layer, respectively, and ε_0 is the permittivity of free space, and ε_r is the permittivity of the dielectric material. For capacitive strain sensors, the gauge factor is the ratio of the relative change in capacitance ($\Delta C/C_0$) to the mechanical strain (ε):

$$GF = \frac{\Delta C/C_0}{\varepsilon} = \frac{1 - \varepsilon \vartheta_e}{1 - \varepsilon \vartheta_d} + \frac{\vartheta_d - \vartheta_e}{1 - \varepsilon \vartheta_d} \tag{4.12}$$

where v_e and v_d are the Poisson coefficients of the electrodes and of the dielectric materials, respectively. GF for capacitive sensors also represents their sensitivity, with higher values indicating higher sensitivity to strain. Nevertheless, the theoretical GF for a parallel plate capacitive strain sensor is 1, regardless of the Poisson ratios (in fact , $\Delta C/C_0 \approx \Delta d/d_0 = \varepsilon$, and from Eq. 4.12, $GF = 1$). This is quite low in comparison to resistive and piezoelectric strain sensors. This GF can be improved by using new sensor configurations or by manipulating the mechanical properties of the material. For example, GF for IDE (interdigitated electrode) of Ag nanowires embedded in a PDMS matrix is $GF = -2$.

The capacitance of a co-planar capacitive sensor is more complex to calculate than parallel plate capacitors due to the fringing fields. However, the capacitance can be approximated based on the geometry and dielectric properties of the material:

$$C \approx \frac{\varepsilon_r l N}{S} \tag{4.13}$$

where N is the number of electrode pairs, l is the overlap length of the fingers, and S is the spacing between them. For high-sensitivity applications, finer electrode patterns and thinner dielectric layers are preferred. A high GF can be achieved by optimizing the electrode design and material properties. For an IDE capacitive strain sensor under strain:

$$\frac{\Delta C}{C_0} = \frac{\Delta l}{l_0} - \frac{\Delta S}{S_0} \tag{4.14}$$

and GF becomes geometry- and material-dependent. It can be >1 or <1, depending upon how these parameters vary under strain.

The sensor's capacitance should be within a measurable range for the intended readout electronics. Typically, capacitance values in the picofarad (pF) to nanofarad (nF) range are suitable. Minimizing parasitic capacitance and ensuring good electrical connections are fundamental for accurate measurements.

4.1.4 Key Parameters in Printed Flexible Strain Sensors

The main performance parameters of printed flexible strain sensors are:

Gauge factor (GF)—is the sensitivity of the strain sensor, being defined as the ratio of relative change in electrical resistance to the mechanical strain. A higher gauge factor indicates a more sensitive sensor.

Stretchability—defines the maximum strain the sensor can withstand before failure. For flexible applications, high stretchability is essential, as it allows the sensor to endure large deformations.

Linearity—measures how consistently the resistance of the sensor changes with applied strain. High linearity is often desirable for accurate readings across the sensor's operational range.

Hysteresis—refers to the sensor's ability to return to its original resistance value after the strain is released. Minimal hysteresis ensures more accurate readings during repetitive loading and unloading.

Response time—is how quick the sensor reacts to changes in strain. Short response times are crucial for applications that involve rapid or dynamic strain changes.

Durability or fatigue resistance—defines the sensor's longevity, particularly its ability to withstand repeated strain cycles without degradation. High durability is essential for applications requiring extended sensor use.

Working range—defines the strain limits within which the sensor operates reliably. This range is determined by the materials and design of the sensor.

Environmental stability—is essential for applications where temperature, humidity, and other environmental factors may affect sensor performance. Materials that offer stability under varying conditions are beneficial.

Size and form factor—The sensor's dimensions, thickness, and overall shape must be optimized for the intended application, ensuring compatibility and ease of integration.

Cost and scalability—For commercial applications, the cost-effectiveness and ease of scaling up the production of sensors are critical, particularly when using printed electronics techniques.

Table 4.1 Comparison of strain sensing approaches (* mV/N)

Sensing type	Sensing mechanism	Gauge factor	Stretchability	Reliability	Linear operating range (%)	Hysteresis
Resistive	Geometric	1–3	10–300%	Very high	10–100	High
Resistive	Percolation	10–100	10–300%	High	5–30	High
Resistive	Crack	100–1000	1–20%	Low	1–10	Very high
Resistive	Disconnection	10–100	1–20%	Medium	1–10	High
Resistive	Tunneling	100–1000	10–200%	High	10–100	Medium
Piezoelectric	Piezoelectric	100–500*	1–5%	High	1–5	Low
Capacitive	Capacitive	1–10	1–50%	Very high	1–30	Low

Each of these parameters plays a critical role in determining the suitability of a printed flexible strain sensor for specific applications.

4.1.5 Comparing Strain Sensing Approaches

Table 4.1 compares the main sensing approaches of flexible printed electrical strain sensors.

4.2 Force and Pressure Sensors

Printed flexible force and pressure sensors measure force or pressure applied to the surface. These sensors are fabricated using printing techniques, enabling them to be thin, lightweight, flexible, and cost-effective. They often rely on piezoresistive, capacitive, piezoelectric and triboelectric effects, where the material's electrical properties change under mechanical stress.

4.2.1 Piezoresistive Mechanism

Piezoresistive mechanism relies on the change of the electrical resistance of a material when subjected to mechanical compressive stress, being able to detect forces or pressures. Main advantages of this type of force and pressure sensors are their simple design, ease of integration with electronic circuits, and high sensitivity. When force or pressure is applied to the sensor, the material deforms, causing a change in its electrical resistive properties. This change in electrical resistance can be measured and correlated to the amount of force or pressure applied. Main sensing mechanisms

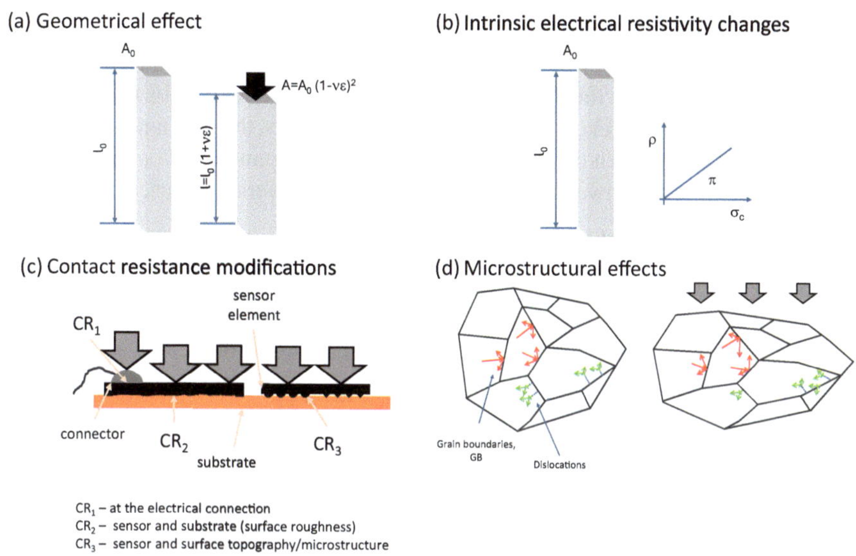

Fig. 4.4 Main force or pressure sensing mechanisms

are based on changes of the electrical resistance due to mechanical deformation, mainly due to geometrical modifications, intrinsic electrical resistivity variations, contact resistance alterations, or microstructural effects. These sensing mechanisms are depicted in Fig. 4.4.

- **Geometrical effect (Fig. 4.4a)**

When a piezoresistive material is deformed upon a compressive stress, its length decreases and the cross-sectional area increases, leading to a decrease in the electrical resistance, according to Eq. (4.1). The electrical resistance, R, of an element of initial length L_0 and cross-area A_0, of a material with a Poisson coefficient of v, when compressed with a strain of ε is of:

$$R = \rho\frac{L}{A} = \rho\frac{L_0(1+\varepsilon)}{A_0(1-\vartheta\varepsilon)^2}. \tag{4.15}$$

Typically, thinner sensing elements tend to have higher sensitivity due to greater strain-induced resistance changes. A longer sensor element increases the electrical resistance, but it may also decrease its sensitivity due to the reduced strain per unit length.

- **Intrinsic electrical resistivity changes (Fig. 4.4b)**

Some materials exhibit changes in their intrinsic electrical resistivity when subjected to a mechanical stress. This is especially common in materials like semiconductors, conductive polymers, or their composites. This relationship can be more complex and depend on the material's composition and structure, as previously seen for strain sensors. In semiconductors, like silicon, deformation causes changes in the electronic band structure, affecting the mobility of charge carriers, which results in a change in the electrical resistivity. In conductive composite materials, the mechanical stress can alter the distance between conductive particles or the alignment of conductive fibers creating conductive pathways, changing the material's overall electrical resistivity.

- **Contact resistance changes (Fig. 4.4c)**

The contact resistance, R_c, can be quantified by the modified Holm contact resistance equation that relates it with the material electrical resistivity, ρ_c, the material hardness, H, the applied normal force, F, and the surface cleanliness coefficient, φ:

$$R_c = \frac{\rho_c}{2 \cdot a} = \frac{\rho_c \cdot \sqrt{\pi \cdot \varphi \cdot H}}{2 \cdot \sqrt{F}} \tag{4.16}$$

where a is the radius of the contact area (assumed to be circular and $a = \sqrt{\frac{F}{\pi H}}$), and φ is an empirical factor ($\varphi = 1$ for clean surfaces). Lower resistivity materials will have lower contact resistance. Increasing the applied load increases the real contact area, reducing R_c, but an excessive force may cause mechanical damage. Softer materials (lower H) result in a larger real contact area under the same load, reducing R_c. A cleaner, smoother surface also reduces R_c. The occurrence of a highly resistive film on the electrical contacts, its corrosion, or contamination leads to $\varphi < 1$.

Figure 4.5 shows main contact resistance sensing mechanisms in flexible force/pressure sensors. In these sensors, R_c is the electrical resistance between the sensing element and its electrical connections, and between the sensing element and the substrate surface. In certain sensor designs, mechanical stress changes the contact resistance between different layers or particles within the sensor. In granular materials or composites, the contact resistance between individual particles can change under pressure, altering the total resistance (interparticle contact). In multilayer sensors, pressure can change the contact resistance at the interfaces between layers (layered structures). For flexible sensors, the contact area can change dynamically with deformation, and consequently the contact resistance and hence the global electrical response.

The contact resistance calculated from Eq. 4.16 is only good for a single asperity. For contact between rough surfaces, R_c can be given by:

$$R_c = \frac{\rho_c}{2 \cdot n \cdot a_c} \tag{4.17}$$

where n is the number of asperity contacts of radius a_c. The electrical resistivity at the interface can significantly impact signal integrity, power efficiency, and the reliability of the sensor. It is affected by the electrical and mechanical properties of the materials, the roughness of their surfaces (at micro and macro levels), possible oxidation or contaminants at the interface, and the contact pressure. With increased

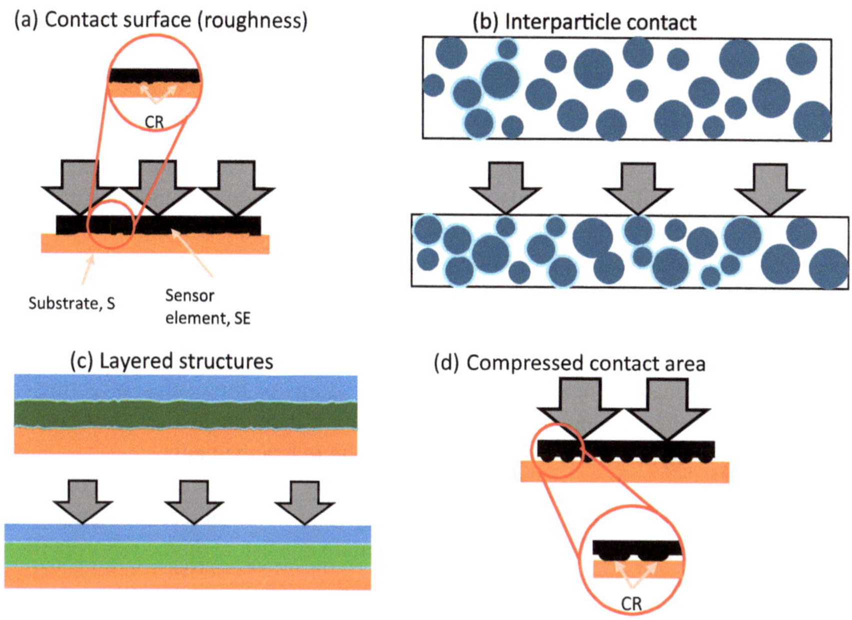

Fig. 4.5 Contact resistance sensing mechanisms

pressure, the number of contact points, n, and/or the contact area, A_c, increases thus reducing R_c. For an elastic deformation (Hertzian contact), A_c is given by:

$$A_c = \pi \cdot \left(\frac{3 \cdot F \cdot r \cdot_{\rho}}{4 \cdot E^*} \right)^{2/3} \tag{4.18}$$

where a is the contact area radius, F is the applied normal force, r_ρ is the effective radius of curvature of the contacting surfaces, and E* is the effective modulus of elasticity (for materials 1 and 2 in a series configuration, $\frac{1}{E^*} = \frac{1-\vartheta_1^2}{E_1} + \frac{1-\vartheta_2^2}{E_2}$, where E_i is Young's modulus and v_i Poisson's coefficient of each material). For more softer materials, the relationships between force and contact area can be established by computer simulations (e.g., FEA).

R_c is affected by the roughness of the surfaces (higher roughness increases R_c) and the presence of oxide layers or contaminants (also increasing R_c). The contact area affects the signal integrity, introducing noise. It can also cause localized heating by Joule effect, affecting sensor stability and performance. Variable contact resistance can lead to nonlinearities in the sensor's output, complicating calibration and reducing reliability. Also, the contact resistance can increase over time due to wear, oxidation, and accumulation of contaminants, leading to sensor drift and malfunction.

- **Microstructural effects (Fig. 4.4d)**

Microstructural changes within the material, such as the reorientation of grains or the movement of dislocations, can affect their electrical properties. In piezoresistive materials, the crystallographic orientation of grains (texture) can significantly influence the piezoresistive coefficients, affecting the sensor's response to mechanical stress. In polycrystalline materials, stress can affect the grain boundaries creating dislocations, increasing the density of grain boundaries, and changing the electrical resistance. Higher density of grain boundaries can increase electrical resistivity due to increased scattering of charge carriers at grain boundaries, affecting the sensor's sensitivity. Also, dislocations can scatter charge carriers increasing electrical resistivity and potentially affecting the sensor's performance. In amorphous materials, deformation can lead to local rearrangements of atoms or molecules, affecting the pathways available for charge transport. This can either increase or decrease the electrical resistivity, depending on the nature of the rearrangements.

The use of porous dielectric materials can exhibit variable dielectric constants, affecting the sensor's accuracy. Materials with high and stable dielectric constants are preferred. Microstructural uniformity helps maintain consistent dielectric properties.

4.2.2 Capacitive Mechanism

The capacitive mechanism is based on the change of the electrical capacitance with deformation, i.e., based on the change in capacitance between two conductive electrodes separated by a dielectric material when pressure is applied. According to Eq. (4.11), the applied force/pressure changes the distance between conductive electrodes (d) or their area ($l \cdot w$), resulting in a change in capacitance. Also, the deformation can alter the dielectric constant (ϵ_r), for compressible dielectric materials.

Three main configurations are normally used for capacitive force/pressure sensors:

- Parallel Plate Capacitors—two parallel conductive plates with a dielectric material in between; pressure reduces the distance between the plates, increasing capacitance.
- Interdigitated Capacitors—interlocking fingers of conductive material on a substrate; pressure changes the overlap area A of the interdigitated fingers or the distance between them, altering the capacitance.
- Variable Area Capacitors—similar to parallel plate capacitors but designed so that the effective area changes with deformation; pressure changes the contact area between conductive surfaces, affecting capacitance.

The conductive layers can be made of metals (e.g., Al, Au), conductive polymers, or printed conductive inks. The dielectric materials generally include flexible polymers, such as PDMS (polydimethylsiloxane), PVC (polyvinyl chloride), TPU (thermoplastic polyurethane), or air gaps in some designs.

4.2.3 *Piezoelectric Mechanism*

A piezoelectric material generates electric charges in response to a mechanical deformation. When a force is applied to a piezoelectric material, the induced deformation leads to a displacement of charges within the crystal lattice of the material, unbalancing them and generating a potential difference (voltage) across it. The generated voltage is proportional to the applied mechanical stress. This phenomenon is used to measure force, pressure, and even dynamic events such as vibrations and accelerations.

The charge generated, Q, by a piezoelectric material under an applied force, F, can be expressed as:

$$Q = d_{ij}F \tag{4.19}$$

where d_{ij} is the piezoelectric charge coefficient, indicating the material's sensitivity to the mechanical stress. The core of a piezoelectric sensor is the piezoelectric material itself, which can be configured in various forms:

a. thin layers of piezoelectric materials deposited on substrates;
b. bulk crystals (single crystals or polycrystalline ceramics);
c. composite structures, combining piezoelectric materials with other materials.

Main piezoelectric materials include natural crystals (e.g., quartz, Rochelle salt), synthetic ceramics (lead zirconate titanate, PZT; barium titanate, $BaTiO_3$) and polymers (polyvinylidene fluoride, PVdF, and their composites). Electrodes attached to the surface of a piezoelectric material collect the generated charges. A piezoelectric ink can be printed and then poled to have a piezoelectric effect (e.g., by a high-voltage discharge, corona poling). Main advantages of these sensors are their high sensitivity to dynamic pressure changes, and their self-powered operation.

The resonant frequency, f, is the frequency at which a piezoelectric structure naturally vibrates with maximum amplitude. At this frequency, mechanical resonance and electrical response are aligned, making the material most efficient at converting electrical to mechanical energy (or vice versa). f of a piezoelectric sensor affects its sensitivity and response time. High-frequency sensors are typically more sensitive. f is calculated from:

$$f = \frac{1}{2\pi}\sqrt{\frac{k}{m}} \tag{4.20}$$

where k is the stiffness of the material, and m is the mass of the sensor ($m =$ density. volume). For a piezoelectric element vibrating in thickness mode:

$$f = \frac{\vartheta}{2t} \tag{4.21}$$

where v is the acoustic wave velocity in the material ($\vartheta = \sqrt{k/\rho}$), and t the piezoelement thickness.

The damping factor measures the rate at which oscillations decay, affecting the sensor's response time and signal clarity. Its value should be optimized for the intended application to balance sensitivity and response time. Sensor damping is determined by:

i. Piezoelectric materials properties (e.g., intrinsic damping)
a high-quality factor (Q) indicates lower intrinsic damping, leading to sustained oscillations, as is the case of piezoelectric polymers like PVdF; Q is also affected by the properties of the substrate material (e.g., its elasticity and internal friction, which also contribute to damping; a soft substrate can introduce mechanical damping, reducing the overall Q of the system); acoustic impedance matching between the piezoelectric material and the substrate can cause reflection and scattering of waves, contributing to energy losses, lowering Q.
ii. Electrode configuration, namely its electrical conductivity and mass
heavier or more conductive electrode materials can add mass and introduce additional damping; thin, lightweight electrodes minimize this effect; complex electrode patterns, like IDE, can influence the distribution of stress and strain, potentially affecting the damping characteristics.
iii. Device geometry
thicker layers can store more energy but may also introduce more internal losses, affecting the damping factor; the shape and size of the sensor can influence how energy is distributed and dissipated, affecting the damping factor; sharp edges can introduce localized stress concentrations that contribute to damping.
iv. Environmental conditions
higher temperatures can increase material damping due to increased internal friction and viscoelastic effects in both the piezoelectric material and the substrate; humidity can be absorbed by the sensor material, particularly polymers, leading to increased damping due to additional mass and altered mechanical properties.
v. Mechanical loading and boundary conditions
the type and magnitude of external mechanical loading (e.g., static or dynamic forces) can affect the damping factor; high stress levels can lead to nonlinear damping behavior; the boundary conditions, i.e., how the sensor is mounted or supported in its application environment, can influence damping; rigid mounting can reduce damping, while flexible or loose mounting can increase it.
vi. Surface conditions
surface roughness and imperfections can scatter acoustic waves, leading to additional energy dissipation and increased damping; additional layers or protective coatings applied for protection or functionalization can contribute to the overall damping of the sensor.

4.2.4 Triboelectric Mechanism

The main principle of this mechanism relies on the generation of an electric charge through friction between two different materials. It involves generating electrical charges through contact and separation of different materials, resulting in the transfer of electrons from one material to the other. This process generates a surface charge on both materials, with one becoming positively charged and the other negatively charged.

When two triboelectric materials are brought into contact, electrons transfer from one material to the other based on their positions in the triboelectric series. Upon separation, the materials retain opposite charges, creating a potential difference (voltage) between them. The relative motion (contact and separation) induces charges on the surfaces of the materials. This can be due to direct physical contact or mechanical deformation, such as pressing and releasing. The separate charges generate an electric field between the materials, which can be captured as an electrical signal. The generated voltage or current can be measured and correlated to the applied force or pressure. This is done through electrodes connected to the triboelectric materials, which capture the induced charge.

Triboelectric materials are chosen based on their position in the triboelectric series to maximize charge transfer. Various combinations of polymers and metals exhibit triboelectric properties, common pairs including PTFE with aluminum or silicone rubber with fabric. They are positioned in a layered structure: typically, a sensor has layers of triboelectric materials separated by a thin dielectric layer to enhance charge separation and collection.

Main advantages of triboelectric force sensors are their high sensitivity, their simple construction, and cost-effectiveness. They have dual functionality in both sensing and energy harvesting, making them useful for self-powered sensor systems. Main limitations of triboelectric force sensors are: (a) their environmental sensitivity (the performance can be affected by humidity, temperature, and surface contamination), (b) their limited durability, as repeated contact and separation can lead to wear and degradation of the materials affecting long-term performance, (c) potential charge leakage and signal drift, as triboelectric charges can dissipate over time, leading to signal drift, (d) they require high-impedance measurement systems, adding complexity and cost to the signal acquisition system.

4.2.5 Key Parameters in Printed Flexible Force and Pressure Sensors

The main performance parameters of printed flexible force and pressure sensors are:

- **Sensitivity**—defines the sensor's ability to detect small changes in force or pressure. Higher sensitivity is beneficial for applications requiring detection of subtle variations.

- **Dynamic Range**—is the range of force or pressure that the sensor can accurately measure. For flexible sensors, a wide dynamic range allows measurement across light and heavy forces, making the sensor versatile for various applications.
- **Repeatability**—refers to the sensor's ability to provide consistent output readings under the same applied force or pressure over multiple cycles. High repeatability ensures reliable data over repeated usage.
- **Linearity**—measures how consistently the sensor output correlates to the applied force or pressure across its working range. A linear response simplifies data interpretation and is desirable for accuracy.
- **Hysteresis**—is the difference in the sensor output when loading and unloading force or pressure. Low hysteresis is essential for applications requiring precise measurements, as it minimizes errors during repeated loading cycles.
- **Response time**—is the time it takes for the sensor to react to a change in force or pressure. Short response times are necessary for applications requiring real-time force or pressure monitoring.
- **Spatial resolution**—refers to the sensor's ability to distinguish between closely spaced points of contact. It is critical in pressure sensors, particularly in applications that need high-detail mapping of pressure distribution.
- **Durability and fatigue resistance**—measure how well the sensor withstands repeated forces over time without degradation. This parameter is important for applications that involve frequent or continuous force measurement.
- **Temperature stability**—is essential for ensuring the sensor's accuracy under varying environmental conditions. Materials that resist temperature-induced changes in electrical properties enhance sensor reliability.
- **Thickness and flexibility**—impact how well the sensor conforms to curved surfaces or integrates into flexible devices. Thinner sensors typically offer greater flexibility and comfort in wearables or robotics.
- **Cost and manufacturing scalability**—are crucial for commercial viability, especially for high-volume applications like consumer electronics or medical wearables.

4.2.6 Comparing Force and Pressure Sensing Approaches

Table 4.2 compares the main sensing approaches of flexible printed force and pressure sensors.

Table 4.2 Comparison of force and pressure sensing approaches

Sensing mechanism	Principle	Materials	Advantages	Common applications
Piezoresistive	Change in resistance with applied force	Conductive polymers	Simple, high sensitivity	Health monitors, tactile sensors, touch screens
Capacitive	Change in capacitance with applied force	Conductive inks (Al, Ag, Au), conductive polymers, flexible dielectric materials (PDMS, PVC, TPU)	High resolution, robust	Touchpads, displays, input devices
Piezoelectric	Electric charge generation with deformation	Piezoelectric ceramics (PZT, $BaTiO_3$) Piezoelectric polymers (PVdF and copolymers)	Self-powered, dynamic sensitivity	Impact sensors, acoustic sensors
Triboelectric	Electric charge generation through friction	PA, PTFE, PVC, PI, PDMS, PANI, PEDOT: PSS	Simple, energy harvesting	Self-powered sensors, wearables, monitoring

4.3 Proximity Sensors

Printed flexible proximity sensors are fabricated on flexible substrates, allowing them to detect the presence or distance of nearby objects without physical contact. Their sensing mechanisms primarily rely on detecting changes in electrical or electromagnetic properties when an object approaches. This can be done based on specific technologies, such as capacitive, inductive, resistive, or optical methods. The specific pattern and design of the printed elements determine the sensor's sensitivity and range. To improve sensitivity, some designs combine printed flexible electronics with thin-film transistors (TFT) or organic electronics. This enhances performance in detecting proximity, making them suitable for more demanding applications.

Each sensing mechanism is chosen based on the specific application needs, environmental factors, and desired accuracy, allowing for versatile and responsive designs of printed flexible proximity sensors.

4.3.1 Capacitive Proximity Sensing

Capacitive sensors measure changes in capacitance that occur when a conductive or dielectric object comes close to the sensor surface. The proximity of the object alters the electric field around the sensor, causing a detectable change in capacitance

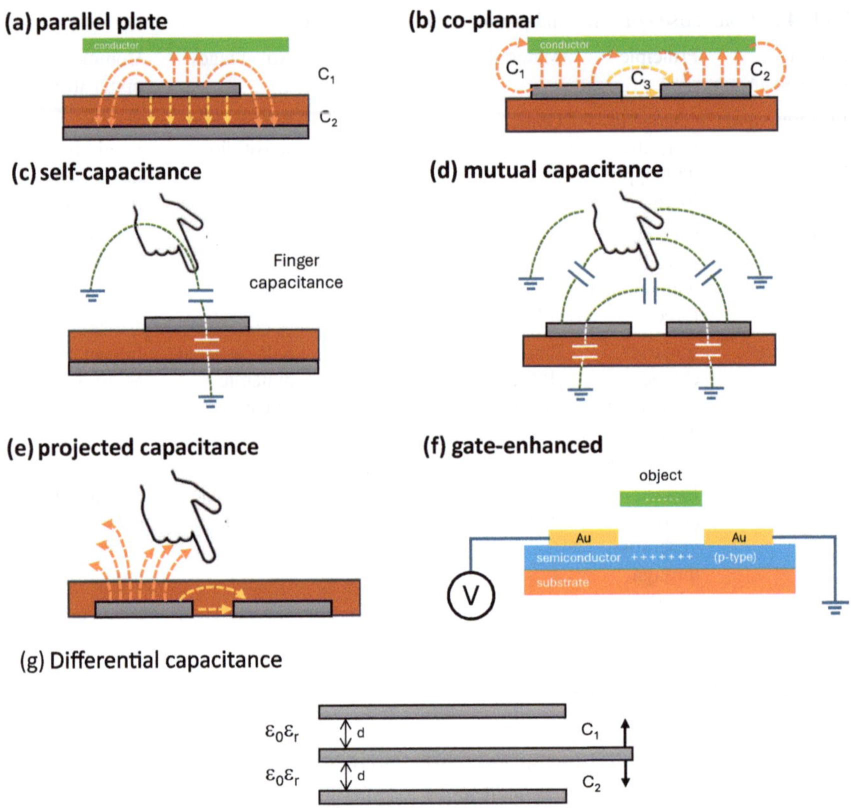

Fig. 4.6 Main types of capacitive proximity sensors based on printed flexible technology

between the sensor's electrodes. Often, they use conductive inks or films on flexible substrates like PET or TPU, with electrode patterns designed to maximize sensitivity.

They can detect a wide range of objects, including non-metallic items like polymers or human skin (e.g., fingers). They are highly sensitivity, have a low-power consumption, and are suitability for various materials (metal, water-based substances, or human skin).

Printed flexible capacitive proximity sensors are designed to be thin, lightweight, and capable of detecting proximity in a range of applications, from wearables to interactive displays. Figure 4.6 shows the main types of capacitive proximity sensors adopting printed flexible electronics technologies.

In capacitive proximity sensors, the variations of the capacitance with approximation of an object are given by Eq. 4.11. The measured capacitance of sensor is positively proportional to the overlap area ($w{\cdot}l$) and inversely proportional to the object distance (d). There are several types of capacitive proximity sensors.

- **Parallel plates and co-planar capacitive sensors (Fig. 4.6a and b)**

In parallel plates capacitive proximity sensors, the electrodes are stacked with a dielectric material in between. The sensor's electric field extends perpendicularly to the plane (C1), detecting objects that approach from above, generating a strong, focused electric field above the sensor surface. When an object approaches, it alters the electrical field and changes the capacitance, which is then detected. In co-planar proximity sensors, the electrodes are arranged side-by-side in the same plane, rather than layered or stacked. The electric field projects horizontally (C3), extending outward along the surface of the substrate. They detect objects approaching from the side or across the plane rather than directly from above. This layout is beneficial for detecting objects that may not be directly above but are close to the sensor's surface.

- **Self-capacitance Sensors (Fig. 4.6c)**

Self-capacitance sensors measure the sensor's own capacitance, formed by a single printed electrode and the approaching object that forms the second "virtual" electrode. When an object approaches, it alters the electric field and the sensor's capacitance, which is then detected. This type of capacitive proximity sensor has a simple design, with only one electrode, and they are sensitive to objects, especially conductive materials.

- **Mutual Capacitance Sensors (Fig. 4.6d)**

Mutual capacitance sensors use two co-planar electrodes to measure capacitance between them. When an object or finger comes close, it changes the capacitance between the electrodes, signaling proximity. They have high accuracy and stability compared to self-capacitance; they are less sensitive to electrical noise and better for multitouch applications.

- **Projected Capacitance Sensors (Fig. 4.6e)**

These capacitive sensors project an electric field from electrodes beneath the surface of a sensor (e.g., a touchscreen). An approaching object changes the field, affecting capacitance and allowing precise location detection. These proximity sensors support high-resolution proximity detection and multitouch functionality and are suitable for thicker cover materials (like glass).

- **Gate-Enhanced Sensors (Fig. 4.6f)**

These sensors are manufactured based on semiconductor materials, and they use the gate electrode to modulate the sensing capabilities and improve sensitivity, selectivity, and control, as in field-effect transistors. These sensors leverage a gate electrode to influence the electrical properties (like capacitance or conductance) of the semiconductor material, creating an electric field that affects the charge carriers, dramatically changing the flowing current. When an object approaches, it alters the capacitance between the gate and the sensing electrode, allowing for detection even at small distances. In gate-enhanced proximity sensors, the air acts as the insulating layer. Typical semiconductors materials are organic polymers, e.g., P3HT (poly(3-hexylthiophene)), TIPS-Pentacene, and PTAA (polytriarylamine).

- **Differential Capacitive Sensors (Fig. 4.6g)**

These sensors use a pair of sensors to measure changes in capacitance difference between them. This setup reduces sensitivity to environmental factors like temperature, making the sensor more accurate. They are highly stable and resistant to interference, suitable for challenging environments.

4.3.2 Resistive Proximity Sensing

Resistive proximity sensors detect changes in electrical resistance due to object approximation. As an object approaches, it can alter the resistance of the printed circuit, which can be measured as a signal indicating distance or presence (Fig. 4.7).

Flexible substrates can be printed/coated with carbon-based materials (carbon black, CNT, graphene), intrinsically conductive polymers (PPy, PANI, PEDOT: PSS), conductive polymers or elastomers (filled with metal particles, Ag, Cu, or C particles), whose resistance changes based on proximity or minor physical deformations. Resistive proximity sensors are low-cost, simple to manufacture, have simple design, and are effective for applications requiring close-range detection.

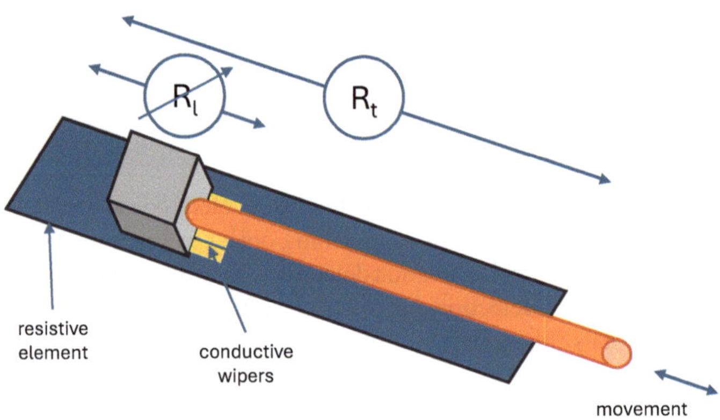

Fig. 4.7 Resistive proximity sensors

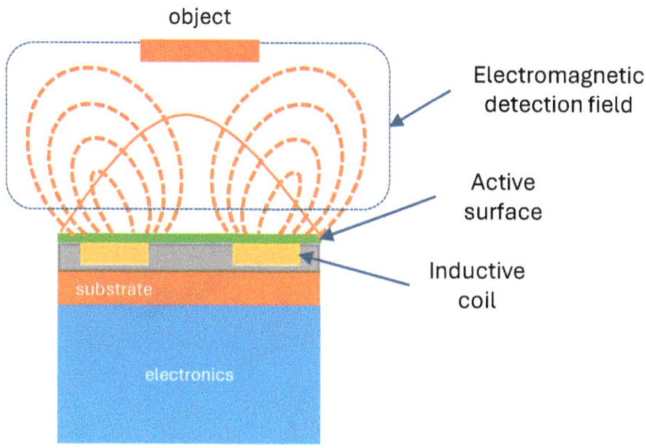

Fig. 4.8 Inductive proximity sensors

4.3.3 Inductive Proximity Sensing

Inductive sensors detect changes in inductance caused by nearby metallic objects. The sensor typically contains a coil, and when a metallic object approaches, it alters the magnetic field generated by the coil (inducing Eddy currents in the object), resulting in a change in the coil inductance that can be measured (Fig. 4.8), indicating proximity. The voltage across the inductor, with inductance L, is given by:

$$V = L\frac{di}{dt} \tag{4.22}$$

where *di/dt* is the time derivative of the current flowing across the inductor.

Flexible substrates are printed with metallic coils, often using Ag-, Cu-, or C-based conductive inks. These proximity sensors are highly accurate with metal objects, are contactless, and immune to environmental contaminants (like dust, dirt, water) and to non-metallic interferences.

4.3.4 Optical Proximity Sensing

Optical proximity sensors work by emitting light (usually IR) and detecting its reflection. When an object comes near, it reflects light back to a photodetector, which generates a signal based on the distance. Organic photodiodes and light-emitting components are printed on flexible substrates, often using organic semiconductors for flexibility. They have an accurate detection over variable distances. They can detect a variety of objects without direct contact. They are highly sensitive and can be used in various lighting conditions. They provide more accurate distance measurements.

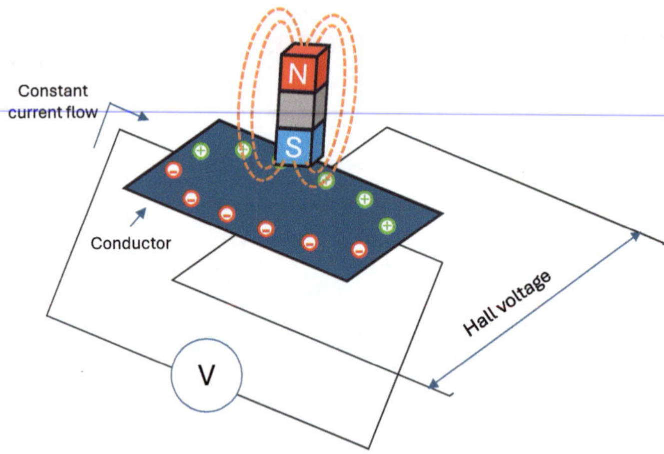

Fig. 4.9 Hall effect proximity sensors

4.3.5 *Magnetic Proximity Sensing (Hall Effect Sensor)*

Hall effect sensors detect changes in a magnetic field caused by the proximity of a magnetic object (Fig. 4.9). When a magnetized object comes near, the magnetic field interacts with the sensor, generating a voltage change proportional to the distance (Hall voltage).

For printed Hall effect sensors, the materials must be both conductive and sensitive to magnetic fields, while also compatible with flexible, printed electronics manufacturing methods. Typical magnetic-sensitive materials are inorganic semiconductors (e.g., Indium Antimonide, InSb, Gallium Arsenide, GaAs) and graphene, organic semiconductors (P3HT, PTAA), magnetic nanoparticle composites (e.g., magnetic nanoparticles like Fe_3O_4 or Co within a polymeric matrix), conductive inks (metallic nanoparticles, like Fe, Ni, Co; conductive polymers, like PEDOT: PSS). Hall effect proximity sensors are durable, precise, and effective even in challenging environmental conditions.

4.3.6 *Multisensing Approaches*

Some flexible proximity sensors combine several sensing mechanisms to improve accuracy and adapt to complex environments. By merging technologies like triboelectric sensing, capacitive sensing, optical sensing, or ultrasonic sensing, hybrid proximity sensors can address the limitations of individual approaches, making them effective in diverse and challenging environments. For instance, combining capacitive and resistive sensing can enhance detection across different object types or conditions.

4.3.7 Key Parameters in Printed Flexible Proximity Sensors

The main performance parameters of printed flexible proximity sensors are:

- **Sensitivity**—measured as a change in electrical signal per unit distance and defines the sensor's ability to detect proximity changes. Higher sensitivity allows detection of objects at greater distances or with finer resolution. Sensitivity is determined by the sensor design and material properties, with capacitive and optical sensors often providing greater range.
- **Resolution**—is the minimum distance change that the sensors can detect. It determines the smallest detectable object or the level of detail the sensor can distinguish. Higher spatial resolution enables the detection of finer movements or smaller objects.
- **Detection distance**—is the maximum distance that the sensor reliably detects the presence of an object. The range is influenced by sensor materials, geometry, and applied voltage.
- **Response time**—is the time taken for the sensor to detect and respond to a change in the proximity of an object. Lower response times are essential for real-time applications, like interactive or motion-sensing systems.
- **Repeatability and stability**—are the ability of a sensor to provide consistent and reliable measurements, i.e., the same output under identical conditions across multiple measurements. High repeatability is essential for reliable and consistent performance. Stability refers to the sensor's ability to maintain performance despite changes in environmental conditions like temperature, humidity, and exposure to chemicals.

Printed flexible proximity sensors open up new possibilities for embedding proximity detection in thin, lightweight, and flexible form factors, enhancing their usability and functionality.

4.3.8 Comparing Proximity Sensing Approaches

Table 4.3 compares the main sensing approaches of flexible printed proximity sensors.

4.4 Temperature Sensors

Printed flexible temperature sensors are designed to detect and measure temperature changes. Typically, they use materials whose resistance changes with temperature. But temperature sensors can operate based on various principles (Fig. 4.10), including resistance changes in thermistors or Resistance Temperature Detectors

Table 4.3 Comparison of proximity sensing approaches

Sensing mechanism	Principle	Materials	Advantages	Common applications
Capacitive	Measures changes in capacitance when a conductive or dielectric object approaches	Conductive inks on flexible substrates like PET, TPU, or paper, dielectrics	High sensitivity, to a range of materials (including non-metals); low-power consumption	Wearable electronics, touchless gesture control, smart devices
Inductive	Detects changes in inductance caused by nearby metal objects altering the magnetic field	Printed metallic coils (Ag or Cu) on flexible substrates	Accurate with metals, immune to dust and dirt interference	Robotics, automotive safety systems, industrial metal detection
Resistive	Measures resistance changes due to object proximity	Conductive polymers, resistive inks, elastomers	Low-cost, simple design; close-range detection	Electronic textiles, interactive displays, close-range sensors
Gate-enhanced	Field effect sensing	Semiconductors, conductive inks	High sensitivity; no contact needed	Touch screens, capacitive proximity sensors
Optical	Uses infrared light emission and reflection to detect nearby objects	Organic photodiodes, printed light-emitting components	Accurate detection over variable distances, versatile	Gesture control in smartphones, touchless interfaces, robotics
Magnetic (hall effect)	Detects changes in the magnetic field from nearby magnetic materials or magnets	Printed magnetic sensors, flexible conductive materials	Reliable for magnetic objects, durable, works under various environmental conditions	Automotive systems, robotics, wearable tech with magnetic elements

(RTD), voltage changes in thermocouples, and changes in material properties like color or fluorescence with temperature. These sensors are lightweight, flexible, adaptable to various surfaces and environments, cost-effective, and easily integrated into various substrates and form factors.

4.4.1 Thermocouples

Thermocouples generate a voltage based on the Seebeck effect, where a junction of two different metals produces a voltage difference, ΔV, related to the temperature

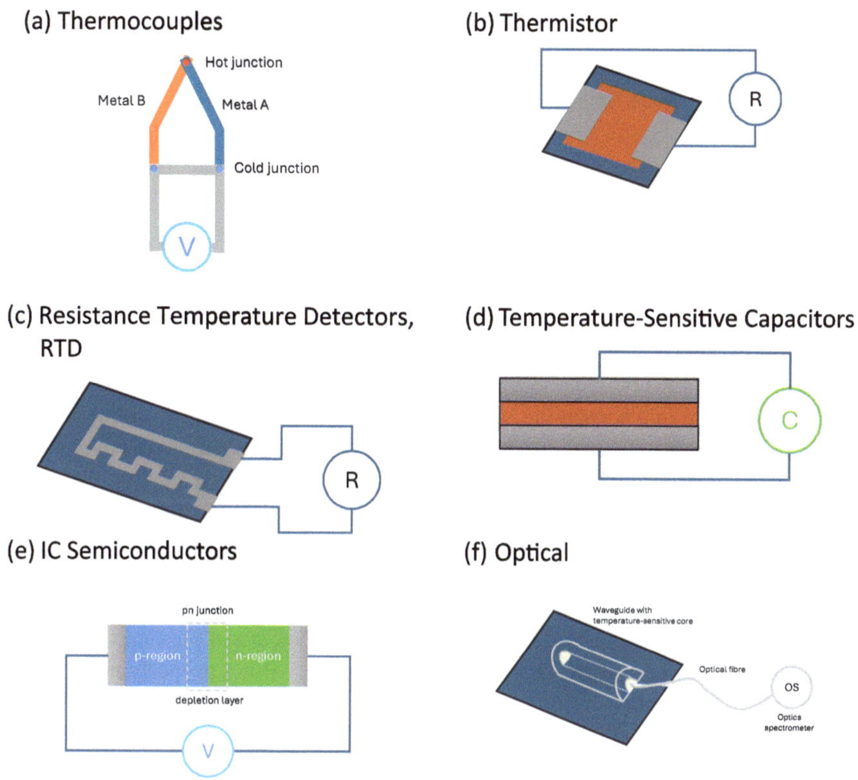

Fig. 4.10 Temperature sensing mechanisms

difference between the hot junction, T_h, and a reference cold point, T_c.

$$\Delta V = S(T_h - T_c) \tag{4.23}$$

where S is the Seebeck coefficient for the selected pair of metals. Thermocouples present a wide temperature range (e.g., from cryogenic levels to extremely high temperatures, up to 1800 °C), are robust and can withstand harsh environments, show a fast response providing real-time temperature monitoring, have simple construction and ease of use with minimal calibration requirements. Thermocouples exhibit nonlinear voltage-temperature relationships, requiring calibration and compensation for accurate measurements. They are less accurate compared to other temperature sensors, typically ranging from ± 1 to ± 2 °C. Thermocouples require compensation for the temperature at the cold junction to provide accurate readings. They can experience drift over time, especially at high temperatures, affecting long-term stability.

Printed thermocouples involve printing two different metal traces on a flexible substrate, joined at one end to form the sensing junction. Bimetallic combinations include, e.g., Copper-Constantan, Nickel–Chromium/Nickel-Aluminum.

4.4.2 Resistance Temperature Detectors, RTD

RTD operate based on the principle that the electrical resistance of certain materials changes predictably with temperature. Platinum, Pt, is commonly used due to its linear, stable, and reproducible resistance–temperature relationship and high accuracy. Other materials include Ni and Cu. Typically, a thin film of Pt is printed onto a flexible substrate in a serpentine pattern to increase the path length and sensitivity. RTD are highly accurate and provide precise temperature measurements over a wide range. They offer excellent long-term stability and repeatability, making them reliable for critical applications. The resistance–temperature relationship of RTD is more linear compared to thermocouples, simplifying calibration and measurement.

RTD, especially platinum-based ones, are more expensive than thermocouples and thermistors. Generally, they show slower response time compared to thermocouples and thermistors, particularly for wire-wound RTD. The current passing through the RTD can cause self-heating, which may affect accuracy if not properly managed.

4.4.3 Thermistors

Thermistors are temperature-sensitive semiconductor resistors whose electrical resistance changes significantly with temperature. In a first-order approximation, it can be assumed that the relationship between electrical resistance, R, and temperature, T, variations is linear:

$$\Delta R = k \cdot \Delta T \tag{4.24}$$

where k is the temperature coefficient of resistance. Instead of k, sometimes the temperature coefficient of resistance, α_T, is used:

$$\alpha_T = \frac{1}{R(T)} \frac{dR}{dT}. \tag{4.25}$$

Negative Temperature Coefficient (NTC) thermistors resistance decrease with increasing temperature, while Positive Temperature Coefficient (PTC) thermistors increase in resistance. In NTC thermistors, the relationship between resistance and temperature is typically exponential. Common metal oxides include oxides of nickel, manganese, cobalt, chromium, and iron (NiO, Ni_2O_3, MnO_2, CrO, Cr_2O_3, CoO). To enhance the temperature sensitivity and achieve the desired electrical properties, metal oxides are doped with small amounts of other materials. The specific dopants used can vary, but they typically include elements like Al, Fe, or other transition metals. NTC thermistors are highly sensitive and precise; they operate effectively over a broad range of temperatures and are very small, compact, and easy to integrate into various devices. Nevertheless, they show a nonlinear response, requiring complex calibration and compensation for precise measurements, and self-heating can cause errors in temperature measurement if not properly managed. NTC thermistors are very sensitive to temperature changes, making them suitable for precise temperature measurements. Many PTC thermistors exhibit a sharp increase in electrical resistance at a specific temperature, known as the Curie point or switching temperature. Doped barium titanate ($BaTiO_3$) shows a sharp increase in resistance at a specific temperature (typically around 120–150 °C, depending on the doping elements). This change is due to a structural transition from a tetragonal (ferroelectric) to a cubic (paraelectric) phase, disrupting the charge carrier movement. PTC polymers consist of conductive particles (e.g., C allotropes) embedded in a polymer matrix. At low temperatures, conductive pathways are formed through carbon particles. As temperature increases, the polymer reaches a critical switching temperature (glass transition temperature), at which it undergoes a high thermal expansion that separates the conductive particles, causing a sharp increase in electrical resistance (typically 60–120 °C, varying with the polymer type). PTC thermistors are particularly sensitive around their switching temperature, making them useful for overcurrent protection. This makes them suitable for overcurrent protection (e.g., circuit breakers), overheat protection (such as in motors and transformers), and self-regulating heaters (heaters and deicing systems where self-regulation of temperature is beneficial). However, they show a limited and narrower temperature range, and lower sensitivity outside their switching region.

4.4.4 Temperature-Sensitive Capacitors

Temperature-sensitive capacitors or capacitive temperature sensors use materials whose dielectric properties (typically the dielectric permittivity, $\varepsilon_r(T)$) change with temperature, affecting the capacitance. In temperature-sensitive capacitors, the dielectric material's permittivity changes with temperature, which in turn affects the capacitance. This change can be used to sense temperature or to compensate for temperature variations in electronic circuits.

A capacitor structure is printed with a temperature-sensitive dielectric material. Dielectric materials can be ceramic (barium titanate, titanium dioxide), but they are highly fragile and support very small deformations. Some polymers also exhibit temperature-dependent permittivity and are used in flexible temperature-sensitive capacitors. Polymers like PVdF, PTFE, PI, PEEK, PET, and PMMA are chosen based on their dielectric properties, thermal stability, mechanical robustness, and ease of fabrication. Temperature-sensitive capacitors can be of PTC (e.g., $BaTiO_3$, certain fluoropolymers and PVdF) or NTC (e.g., PS, PC) type.

Temperature-sensitive capacitors are highly sensitivity to temperature changes, making them effective for precise temperature sensing and compensation. They are durable and reliable with a long operational life, especially when using ceramic materials. Some temperature-sensitive capacitors exhibit nonlinear changes in capacitance with temperature, requiring complex compensation techniques. The effective temperature range may be limited by the properties of the dielectric material. The performance is highly dependent on the quality and characteristics of the dielectric material, which can vary between manufacturers. Proper material selection and circuit integration are decisive for optimizing their performance in specific applications.

4.4.5 Semiconductor Temperature Sensors

Semiconductor temperature sensors or IC-temperature sensor, such as diodes and transistors, rely on the temperature-dependent forward voltage drop across a pn-junction (a combination of two semiconductors of p-type and n-type in a single device; e.g., a p–n junction connected on both sides to a circuit is a diode; a bipolar junction transistor, BJT, is in the form n–p–n or p–n–p). As temperature increases, the forward voltage drop decreases, which can be measured and correlated to temperature changes.

For example, a printed silicon thermistor (PST) can be used in a voltage divider configuration to provide a direct link between output voltage and temperature. Semiconductor-based temperature sensors have good linearity and accuracy and are easily integrated within digital systems, but they are slow to respond to fast temperature changes. These devices can be printed on flexible substrates, making them suitable for wearable applications. They offer high accuracy (within $\pm$ 0.1 °C) and fast response times (less than 100 ms), which are needed for real-time monitoring in applications such as health tracking and environmental sensing. These sensors can cover a large active sensing area and can be integrated into analog circuits or temperature sensor arrays for more complex applications.

Materials like polydimethylsiloxane (PDMS) and various carbon-based materials (e.g., CNT, graphite) are often used as the base materials for these flexible sensors. These materials are chosen for their excellent electrical and mechanical properties, and low manufacturing costs, making the sensors more sensitive and durable.

4.4.6 Optical Temperature Sensors

Optical sensors use materials that change their optical properties (such as fluorescence or absorbance) with temperature. Printed sensors can use these materials to detect temperature changes based on variations in light intensity or wavelength, typically, printed layers of temperature-sensitive dyes or polymers. Optical temperature sensors allow for non-contact measurements, with high sensitivity. Optical sensors are detailed in Sect. 4.6.

4.4.7 Key Parameters in Printed Flexible Temperature Sensors

The main performance parameters of printed flexible temperatures sensors are:

Sensitivity—is the degree of change in the sensor's output per unit change in temperature. High sensitivity enables the sensor to detect small temperature variations, which is important for applications requiring precise monitoring.

Response time—is how quickly the sensor can detect a temperature change. Fast response times are important for applications that involve rapid temperature fluctuations, enabling real-time monitoring.

Temperature range—defines the minimum and maximum temperatures the sensor can accurately measure. For broader applications, sensors with a wide range are desirable, particularly if the environment includes extreme temperatures.

Linearity—indicates how consistent the sensor's output is across its operational temperature range. High linearity simplifies calibration and improves accuracy, especially when a straightforward relationship between temperature and output is required.

Accuracy and precision—accuracy is how close the sensor's measurements are to the true temperatures, while precision measures the consistency upon repeated readings.

Stability and drift—stability is the sensor's ability to maintain accuracy over time, while drift refers to changes in output under constant temperature conditions. Low drift and high stability are key for long-term monitoring applications.

Repeatability—describes the sensor's ability to provide consistent readings over multiple temperature cycles. This parameter is essential for sensors expected to endure repeated heating and cooling.

Hysteresis—is the difference in the sensor output during heating and cooling cycles. Low hysteresis is desirable for applications where temperature changes frequently, as it improves the accuracy and consistency of readings.

Flexibility and mechanical robustness—are critical for sensors that need to be applied to curved or moving surfaces. Mechanical robustness ensures that the sensor can withstand bending, stretching, or deformation without performance degradation.

Environmental stability—ensures that the sensor performs consistently in varying humidity, pressure, and environmental conditions. It's especially critical for sensors used in outdoor or variable environments.

Size and thickness—The physical dimensions of the sensor, including its thickness, influence its application and integration into devices. Thinner sensors with small form factors are advantageous for wearables, medical devices, and flexible electronics.

Cost and scalability—Cost-effectiveness and the potential for large-scale production are fundamental for commercial applications. Printed flexible sensors often benefit from low-cost production methods, making them suitable for disposable or high-volume uses.

4.4.8 Comparing Temperature Sensing Approaches

Table 4.4 compares the main sensing approaches of flexible printed temperature sensors.

4.5 Humidity Sensors

Humidity sensors detect and measure humidity levels in the environment. They can operate based on changes in electrical properties (such as capacitance or resistance) when exposed to humidity. Sensor encapsulation protects it from mechanical damage and environmental factors, while allowing humidity to reach the sensing material. Common encapsulation materials include silicone and polyurethane. Flexible humidity sensors can conform to complex surfaces and maintain performance under bending and stretching. Main challenges on flexible humidity sensors are related to: a) durability and ensuring long-term stability and resistance to mechanical stress, environmental factors, and repeated use; b) achieving high sensitivity and accuracy comparable to traditional rigid sensors, requiring careful material selection and design optimization; c) designing robust signal processing circuits to handle the variable outputs from flexible sensors; and d) ensuring compatibility of the substrate, ink, and sensing materials for reliable performance over a wide temperature and humidity ranges.

4.5.1 Piezoresistive

These sensors measure humidity by detecting changes in the electrical resistance due to the adsorption of water vapor on a hygroscopic material. As humidity increases, water molecules adsorb onto the surface of the sensing material, changing its electrical resistance. Typically, this involves either increasing or decreasing the electrical conductivity, depending on the material used. Polymers like polyvinyl alcohol (PVA) and polyimide change electrical properties with humidity. Conductive polymers, as PANI, polypyrrole (PPy), PEDOT: PSS and polymer-based nanocomposites (e.g., CNT or graphene mixed with hydrophilic polymers), are used due to their humidity-dependent electrical properties. Metal oxides such as zinc oxide (ZnO), tin dioxide (SnO_2), and titanium dioxide (TiO_2) are also used for their reliable humidity sensing

Table 4.4 Comparison of temperature sensing approaches

Sensing mechanism	Principle	Materials	Advantages	Common applications
Thermocouples	Generate a voltage proportional to temperature difference (Seebeck effect)	Various metal pairs (e.g., type K: chromel–alumel, type J: iron-constantan)	Wide temperature range (−200–1800 °C), durable, fast response	Industrial processes, HVAC systems, automotive
Thermistors	Measure electrical resistance changes with temperature variations	Metal oxides (e.g., MnO_2, NiO, CoO) for NTC; polymers for PTC	High sensitivity, fast response, cost-effective	Consumer electronics, medical devices, automotive, environmental monitoring
Resistance temperature detectors, RTD	Electrical resistance of changes with temperature	Platinum, nickel, copper	High accuracy and precision, excellent stability and repeatability, linear response	industrial process control, HVAC systems, automotive
Temperature-sensitive capacitors	Change capacitance with temperature due to variations in the dielectric constant	Ceramics (e.g., barium titanate), polymers	High sensitivity, non-intrusive, reliable, and durable	RF circuits, oscillators, environmental monitoring, industrial processes
Semiconductor junction-based (IC-integrated circuit)	The forward voltage drop across a semiconductor pn-junction changes with temperature	Silicon-based semiconductors are commonly used	Linear response, ease of integration into digital systems, good accuracy	Consumer electronics, environmental monitoring, automotive sensors, and IC for temperature compensation
Optical	Change of optical properties with temperature	Temperature-sensitive dyes, polymers	Non-contact measurement; high sensitivity	Wearable sensors, flexible displays, medical diagnostics

capabilities, exhibiting changes in conductivity or capacitance with humidity. The humidity sensor is based on patterned conductive paths that are designed to maximize the interaction with the sensing material and to optimize electrical resistance changes.

4.5.2 Capacitive

These sensors measure humidity by detecting changes in capacitance caused by the absorption of water vapor in a hygroscopic (water-absorbing) dielectric material. When humidity increases, water molecules are absorbed by the dielectric material, increasing its dielectric constant and, consequently, the capacitance of the sensor. Polymer-based dielectrics like PI, polyethylene oxide (PEO), cellulose acetate, and PVA are common choices due to their high sensitivity to humidity changes. Nanocomposites, such as graphene oxide (GO) and polymer blends, are also used to improve sensitivity. Thin films of materials like tin oxide (SnO_2) or titanium dioxide (TiO_2) exhibit changes in conductivity or capacitance with humidity. IDE designs are used in capacitive humidity sensors as they maximize the surface area for interaction with the hygroscopic dielectric material, enhancing sensitivity.

4.5.3 Key Parameters in Printed Flexible Humidity Sensors

The main performance parameters of printed flexible humidity sensors are:

Sensitivity—refers to how well the sensor can detect small changes in humidity levels. High sensitivity is essential for precise monitoring.

Response time—is the time the sensor takes to react to a change in humidity. Faster response times allow immediate detection of environmental changes.

Humidity range—specifies the minimum and maximum levels of relative humidity the sensor can accurately measure. A wide range is often required for applications that involve extreme humidity variations.

Linearity—indicates how consistently the sensor's output correlates to changes in humidity across its range. High linearity is desirable for easy calibration and accurate readings.

Accuracy and precision—is how close the sensor's output is to the actual humidity value. Precision reflects the consistency of repeated measurements. High accuracy and precision are essential for reliable data, especially in medical or environmental applications.

Hysteresis—is the difference in the sensor's response during increasing and decreasing humidity cycles. Low hysteresis is essential for applications with frequent humidity changes, as it reduces reading inconsistencies.

Repeatability—is the sensor's ability to consistently produce the same output under identical conditions over multiple humidity cycles. High repeatability ensures reliable long-term monitoring.

Stability and drift—refers to the sensor's ability to maintain consistent readings over time, while drift indicates changes in output under constant conditions. Low drift and high stability are critical for sensors used in applications requiring prolonged monitoring.

Temperature dependency—Since humidity readings can be influenced by temperature, a sensor's temperature dependency should be minimal or well-compensated. This factor is decisive for applications where temperature fluctuates, ensuring accuracy across varying environmental conditions.

Flexibility and mechanical robustness—Flexibility is essential for applications requiring the sensor to conform to curved surfaces or endure bending. Mechanical robustness ensures that the sensor's performance is not affected by the physical deformation, which is critical in wearable and flexible electronics.

Environmental Stability—refers to the sensor's performance under different environmental factors, such as pressure, dust, or chemicals. It's important for applications in harsh or outdoor environments.

Recovery Time—is the period the sensor takes to return to its baseline state after humidity changes. A short recovery time ensures that the sensor is ready for the next measurement quickly, enhancing responsiveness.

Size and thickness—The sensor's physical dimensions, including thickness, impact its application and integration into flexible or compact devices. Thin, small sensors are preferred for wearables and compact devices.

Cost and manufacturing scalability—are critical for commercial viability, particularly for sensors intended for mass-market or disposable applications. Printed electronics techniques often provide a cost-effective approach for large-scale production.

4.5.4 Comparing Humidity Sensing Approaches

Table 4.5 compares the main sensing approaches of flexible printed humidity sensors.

4.6 Optical Sensors

Printed flexible optical sensors are an emerging technology that combines the advantages of optical sensing with the flexibility, lightweight nature, and cost-effectiveness of printed electronics. They are advanced devices that detect changes in light properties, such as intensity, wavelength, phase, or polarization, to measure various physical, chemical, or biological parameters. These sensors measure changes in optical properties of light, such as absorption, fluorescence, surface plasmon resonance, due

Table 4.5 Comparison of humidity sensing approaches

Sensing mechanism	Principle	Materials	Advantages	Common applications
Resistive	Changes in resistance due to humidity adsorption	Conductive polymers (PANI, PPy, PEDOT), metal oxides (ZnO, SnO_2)	High sensitivity Lower costs	Consumer electronics, automotive, industrial, environmental monitoring
Capacitive	Changes in capacitance due to dielectric constant variation	PI, PEO, PVA, cellulose acetate, GO nanocomposites, barium titanate ($BaTiO_3$)	High sensitivity Less affected by temperature More linear response, easier to calibrate Lower power consumption	HVAC systems, industrial control, medical devices, environmental monitoring

to the presence of an analyte. Typically, an optical sensor consists of a light source (e.g., OLED, printed quantum dots), a photodetector (e.g., printed photodiodes), and sometimes a waveguide or photonic crystal structure. Figure 4.11 shows an overview of the layered structure of a typical printed flexible optical sensor.

The structure of an optical sensor comprises:

- Flexible substrate that provides a durable and flexible base for the sensor (e.g., PET, PI).
- Conductive electrode layer forms the electrical connections and pathways (e.g., Ag ink)
- Optical sensing layer that interacts with the analyte and produces optical changes (e.g., PMMA infused with fluorescent dye or chromophore, nanoparticles).
- Encapsulation layer protects the sensor components from environmental damage (e.g., transparent silicone).
- Photodetector layer that detects changes in light properties and converts them into electrical signals (e.g., printed photodiodes).

Light from the light source (not shown in Fig. 4.11) passes through the sensor layers. The optical sensing layer interacts with the analyte, causing changes in fluorescence or other optical properties. The photodetector layer captures the altered

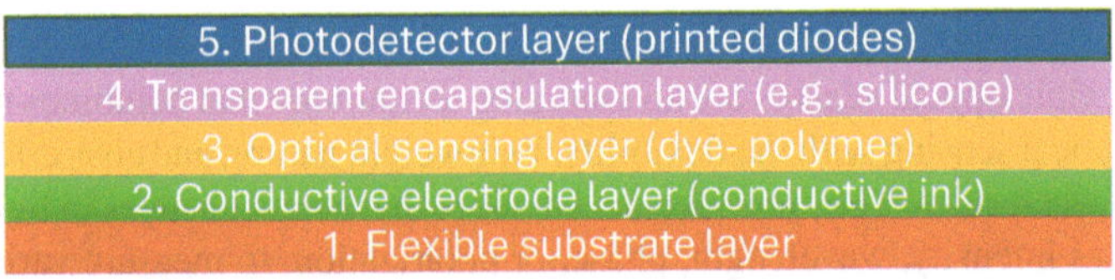

Fig. 4.11 Structure of an optical sensor

light and converts it into an electrical signal for analysis. The precise control of the thickness of each layer ensures proper function and signal integrity. Optical sensors have a high sensitivity and specificity, provide non-invasive measurements, and are capable of multiplexing.

4.6.1 *Optical Sensing Approaches*

Table 4.6 compares the main optical sensing approaches of flexible printed sensors.

4.6.2 *Main Types of Optical Sensors*

Optical sensors can be used to measure different types of optical stimuli. Printed photodetectors measure light intensity and are sensitive to various wavelengths, including UV, visible, and IR. Colorimetric sensors rely on materials that change color in response to specific analytes or environmental conditions, such as pH or the presence of gases. Optical humidity sensors measure changes in relative humidity levels using photonic crystal sensors (that change color with humidity) or waveguide-based humidity sensors. Temperature optical sensors measure changes in temperature, typically by observing IR radiation or using temperature-sensitive materials (e.g., thermochromic materials). Optical proximity sensors detect the proximity or distance of an object, typically by detecting reflected light (e.g., ToF sensors, IR proximity sensors). Optical sensors, as biomarkers, can be used to detect the presence and concentration of specific biomolecules, such as glucose, proteins, or DNA. Changes in the refractive index of a surrounding medium, often indicating chemical or biological interactions, can be detected by optical sensors. Strain, stretching, or bending alters the sensor's structure or their optical properties, which can be measured by optical waveguide strain sensors, fiber optic strain sensors, or photonic crystal strain sensors. The presence of particles (e.g., dust, smoke) or pathogens (e.g., bacteria, viruses) can be based on optical scattering or fluorescence.

4.6.3 *Main Materials*

Main materials for components of optical sensors are presented in Table 4.7.

Sensor geometries are typically based on IDE design that maximizes the interaction area between the electrodes and the gas-sensitive material. Planar waveguides are used in SPR sensors to guide light efficiently and interact with the sensing material. Efficient coupling of light into and out of the sensor must be ensured, through the design of optical waveguides or lenses. Also, light isolation must be guaranteed,

Table 4.6 Comparison of optical sensing approaches

Sensing mechanism	Principle	Operation
Absorption spectroscopy	Measures the decrease in light intensity as it passes through a medium where the analyte absorbs specific wavelengths	A light source emits light through a gas or liquid sample, and the sensor detects the intensity of light before and after it passes through the sample; the reduction in light intensity at specific wavelengths indicates the presence and concentration of the analyte
Fluorescence spectroscopy	Measures the emitted light from a fluorophore-labeled analyte or product upon excitation by a specific wavelength	The analyte is labeled with a fluorescent dye, which emits light when excited by a certain wavelength; the intensity of the emitted light is proportional to the analyte concentration
Surface Plasmon resonance (SPR)	Measures changes in the refractive index near the sensor surface caused by the binding of analytes	A light beam excites surface plasmons on a metal film; the binding of analytes changes the local refractive index, altering the resonance condition, which is detected as a change in the reflected light intensity
Raman spectroscopy	Relies on inelastic scattering of light, where a small fraction of the incident light is scattered at different wavelengths due to interactions with molecular vibrations	Each gas has a unique Raman spectrum, which can be used for identification and quantification; especially those that don't have strong IR absorption, such as N and H
Colorometric sensing	Detects a color change due to the chemical reaction between the dye and gas, altering its light absorption characteristics	A gas-sensitive dye or pigment is printed onto a flexible substrate; when exposed to the target gas, it undergoes a color change that can be detected visually or through an optical detector
Optical waveguide	Measures changes in light propagation through a waveguide caused by the interaction of analytes with the waveguide surface	Analytes binding to the waveguide surface affect the light propagation, which is detected as a change in output light intensity or phase

ensuring minimization of ambient light interference and crosstalk between sensor elements.

4.6.4 Main Features of Optical Sensors

Key benefits of printed optical sensors are various, such as:

Table 4.7 Main materials for components of optical sensors

Component	Materials	Operation
Light sources	Organic LED	Flexible, printable light sources for excitation
	Printed quantum dots	Tunable emission wavelengths and flexibility
Detectors	Printed photodiodes and photodetectors	Convert light into electrical signals, suitable for integration with flexible substrates. Organic semiconductors (P3HT, PCBM), perovskites, quantum dots, carbon-based materials (e.g., graphene, CNT)
	Photomultipliers	Highly sensitive detectors for low-intensity light, used in fluorescence spectroscopy
	CCD/CMOS cameras	Capture images of fluorescence or other optical signals, used in high-resolution applications
Sensing materials	Dye-infused polymers	Fluorescent dyes, PS or PMMA doped with sensitive dyes, organic fluorophores
	Nanomaterials	Graphene oxide, functionalized CNT, quantum dots, Au and Ag nanoparticles
	Photonic polymers	Colloidal crystals or photonic inks
	Colorimetric	Color-changing dyes, metal–organic frameworks (MOF), and responsive polymers that change color upon exposure to certain chemicals or humidity
Optical components	Lenses	UV-curable polymers (acrylate-based resin and epoxies, PDMS, PMMA)
	Mirrors	Ag and Al nanoparticle inks and dielectric coatings (SiO_2 TiO_2) for reflective coatings, polymeric substrates (e.g., PET) with metallic coatings
	Optical fibers	Polymer Optical Fibers (POF) based on PMMA, PC or UV-curable resins
	Prisms	UV-curable transparent polymers, epoxy resins, PMMA
	Gratings	Photoresist polymers (e.g., SU-8), holographic photopolymers, metal nanoparticle inks (Ag, Au)

a. their cost-effective production (e.g., printing techniques reduce manufacturing costs, especially for large-area sensors) makes them suitable for IoT and smart devices;
b. their flexibility, conformability to complex surfaces, lightweight and thin design enhance comfort and usability in wearable and embedded applications;
c. their high sensitivity and specificity allow them to detect low concentrations of specific biomolecules with high accuracy;
d. their scalability and customization can be tailored to specific applications and mass production;
e. their high sensitivity and precision to small changes in stimuli;

f. the rapid response time, providing real-time monitoring and rapid detection of biological interactions; and
g. their environmental and chemical robustness.

But they have main limitations in terms of durability, being difficult to ensure long-term stability and resistance to mechanical stress, environmental factors, and repeated use. They have more complex design and fabrication processes compared to traditional rigid biosensors. They show the potential of interference from other biomolecules and environmental factors, requiring careful calibration and compensation. Depending on the optical components used, power consumption can be higher compared to other types of sensors.

4.6.5 Key Parameters in Printed Flexible Optical Sensors

The main performance parameters of printed flexible optical sensors are:

Sensitivity—refers to the sensor's ability to detect small changes in light intensity. High sensitivity is beneficial in applications requiring precise detection of subtle light changes.

Spectral sensitivity—indicates the range of wavelengths (e.g., UV, visible, IR) the sensor can detect. For applications like environmental monitoring or biomedical analysis, specific wavelength sensitivity is essential.

Responsivity—measures the sensor's output signal per unit of light input (usually in A/W). Higher responsivity improves the sensor's ability to detect low-light levels, which is critical for low-light or high-precision applications.

Response time—is how quickly the sensor reacts to changes in light intensity. Fast response times are essential for applications requiring real-time monitoring or rapid changes in illumination.

Linearity—describes how consistently the sensor's output corresponds to changes in light intensity across its operating range. High linearity is desirable for accurate readings and easier calibration.

Dynamic range—is the range of light intensities the sensor can accurately detect. A wide dynamic range is ideal for applications that may encounter both low and high light intensities, such as outdoor or variable lighting environments.

Signal-to-noise ratio (SNR)—is the level of the desired signal relative to the background noise. A high SNR is essential for accurate detection, especially in low-light conditions where noise can interfere with signal clarity.

Hysteresis—is the difference in the sensor's output between increasing and decreasing light intensities. Low hysteresis is essential for applications with variable lighting conditions, ensuring consistent reading.

Dark current—is the current that flows through the sensor in the absence of light. Low dark current is essential for low-light applications, as it reduces background noise and improves accuracy.

Environmental stability—ensures that the sensor's performance remains consistent under varying temperature, humidity, and environmental conditions. This parameter is critical for outdoor or exposed applications.

Flexibility and mechanical durability—flexibility is essential for applications where the sensor is integrated into curved or moving surfaces, such as wearables or robotic skins. Mechanical durability ensures that the sensor maintains performance under bending or stretching.

Temperature dependency—measures how temperature variations affect the sensor's output. Minimizing temperature dependency or adding compensation mechanisms improves accuracy in fluctuating environments.

Form factor (size and thickness)—the dimensions of the sensor (e.g., thickness) influence its application and integration. Smaller, thinner sensors are preferred for wearables, compact devices, and surfaces with limited space.

Power consumption—low-power consumption is critical for battery-powered or portable devices, where energy efficiency extends operational time and minimizes the need for frequent recharging.

Cost and scalability—are decisive for commercial applications, particularly for mass-market or disposable sensors. PE techniques often offer a low-cost approach suitable for large-scale production.

4.7 Gas and Chemical Sensors

Printed flexible gas sensors are designed to detect and measure the concentration of gases in various environments. Chemical sensors detect specific chemical compounds through interactions that cause measurable changes in electrical properties, mass, or optical signals. Electrochemical sensors detect chemical changes through redox reactions that generate a measurable electric current or potential. These sensors are commonly used for detecting specific gases or chemical compounds, pH levels, and biological analytes. Printed functional inks containing materials that interact with target substances cause measurable changes in electrical properties. IDE sensor design maximizes the interaction between the sensing material and gas molecules. Patterned conductive path optimizes the sensor's response by designing paths that enhance gas-sensing interactions. Encapsulation protects the sensor from mechanical damage and environmental factors. Common encapsulation materials include silicone and polyurethane. These sensors are flexible, lightweight and thin, cost-effective, and easily integrated into various substrates and form factors, expanding their application potential. Their flexibility and conformability enable their integration into a wide array of applications, including wearable electronics, environmental monitoring, medical devices, and smart packaging.

Main challenges on printed flexible gas or chemical sensors are related to: (a) sensor durability, ensuring long-term stability and resistance to mechanical stress, environmental factors, and repeated use; (b) achieving high sensitivity and selectivity for specific gases while minimizing cross-sensitivity to other gases and environmental factors; (c) designing robust signal processing circuits to handle the variable outputs from flexible sensors; and (d) ensuring compatibility of the substrate, ink, and sensing materials for reliable performance over a wide temperature and humidity ranges. Main sensing mechanisms include chemiresistive, capacitive, electrochemical, and optical.

4.7.1 Chemiresistive Sensing

In this type of sensor, its electrical resistance changes when the sensing material interacts with gas molecules. Adsorption or reaction of gas molecules on the surface of the sensing material alters its electrical conductivity. Chemiresistive sensing is a widely used method for detecting gases and volatile organic compounds (VOC) due to its simplicity, high sensitivity, and ease of integration into various devices.

Chemiresistive sensors rely on materials that undergo resistance changes due to adsorption or reaction with target substances, such as (a) conductive polymers (PEDOT: PSS, PANI, PPy, polythiophene derivatives); (b) carbon-based materials (like graphene and GO, rGO, CNT, carbon black); (c) metal oxides (ZnO, SnO_2, TiO_2, Fe_3O_2, NiO); (d) metal nanoparticles and nanocomposites (e.g., Ag, Au, Cu, metal oxides nanoparticles); (e) organic–inorganic hybrids (like perovskites, MOF, zeolites); and (f) composite materials (e.g., polymer-carbon, polymer-metal oxides, perovskite-carbon composites).

4.7.2 Capacitive Sensing

The sensor's capacitance changes when gas molecules or other specific chemicals are absorbed or adsorbed in a dielectric material. Gas/chemical molecules interacting with the dielectric material change its electrical permittivity, affecting the sensor's capacitance. The basic structure consists of a pair of electrodes separated by a gas/chemical sensitive dielectric layer. The dielectric materials play a crucial role in capacitive sensors, as their properties determine how capacitance changes in response to gas or chemical exposure. These materials can absorb or interact with target analytes, resulting in changes in the electrical permittivity. Typical dielectric materials include PVA, PDMS, PI, PEI, PEO, PVA, cellulose nanocrystals. Other functional materials are carbon-based materials, metal oxides nanocomposites, metal nanoparticles nanocomposites, MOF, and organic–inorganic hybrids.

These sensors are highly sensitivity, capable of detecting small changes of gas/chemicals. They show a more linear response compared to resistive sensors, simplifying calibration and improving accuracy. They also require less power to operate, making them suitable for battery-operated and portable devices. Nevertheless, they generally have higher cost compared to resistive humidity sensors due to the complexity of materials and manufacturing processes.

4.7.3 Electrochemical Sensing

An electrochemical sensor generates an electrical signal through an electrochemical reaction between the gas/chemical and the sensing material. Gas/chemical molecules react with the sensing material, causing a change in current or voltage. Main advantages of electrochemical sensors are their high sensitivity and resolution, their specificity to certain gases with reduced cross-sensitivity, their rapid detection and response, suitable for real-time monitoring. Their miniaturization allows the integration into small, portable devices. However, they show more complex design and fabrication processes compared to capacitive or resistive sensors. Their long-term stability can be an issue, with performance degradation over time due to environmental factors and material aging. Furthermore, their sensitivity to temperature variations requires compensation techniques. Generally, they have a higher cost due to the use of noble metal electrodes and complex fabrication processes.

Electrochemical sensors are analytical devices used to detect or measure chemical substances by converting their concentrations into an electrical signal. Common sensor configurations are planar electrodes with two-electrode system (combining working and reference electrodes for simplicity) or three-electrode system (including a working, reference, and counterelectrodes for more accurate and controlled measurements, as depicted in Fig. 4.12). The sensor configuration determines the efficiency of electron transfer, the sensitivity, and the overall performance of the sensor. The effective surface area of the WE determines sensor performance. Larger surface areas enhance sensor sensitivity by providing more active sites for redox reactions. This can be achieved by nanostructuring or modifying the electrodes surface, which can increase the effective surface area. The distance between WE, RE, and CE also affects the efficiency of electrochemical cells and the diffusion of ions. Proper spacing is important for minimizing electrical resistance and maximizing sensitivity.

The conductive materials used for WE, RE, and CE affect the electrical conductivity, the stability, and the specificity of the sensor. The choice of material can also

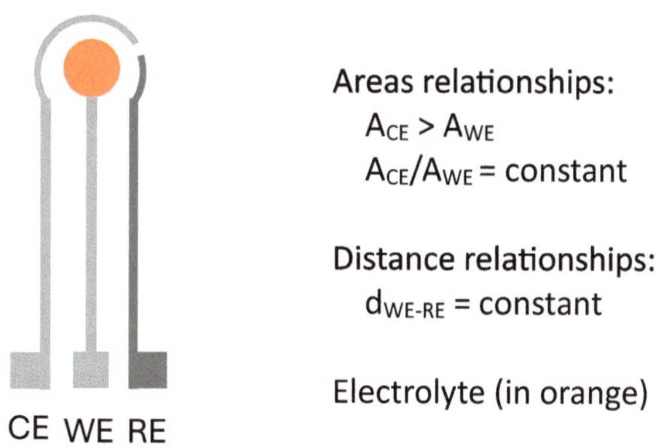

Fig. 4.12 Configuration of an electrochemical sensor and main design parameters

influence the reaction kinetics and electrode surface properties. Typically used materials for WE are Au, Pt, C (graphite, graphene, CNT) due to their electrical conductivity and catalytic properties; Ag/AgCl are selected for a stable potential RE, and, for CE inert materials such as Pt or C. Functionalization of the electrode surface with specific materials can be done to enhance selectivity to specific target analytes and improve sensitivity. Common modifications include enzyme immobilization, metal nanoparticles, conductive polymers, and molecularly imprinted polymers (MIP). The electrolyte, placed over WE, is the ionic medium in which the electrochemical reactions occur, being liquid, gel, or solid-state electrolytes depending on the application. Typical electrolytes are solutions based on phosphate-buffered saline (PBS), potassium chloride (KCl), or sulfuric acid (H_2SO_4). The electrolyte affects electrical conductivity, sensor stability, and reaction kinetics. The pH and ionic strength of the electrolyte can also impact sensor performance. Solid electrolytes can also be used, based on phosphoric acid-doped polymers (e.g., polybenzimidazole, PBI), sulfonated polystyrene (SPS), or perfluorosulfonic acid (PFSA)/polytetrafluoroethylene (PTFE) copolymers (Nafion®), due to their proton conductivity and stability. A selective membrane, or separator, separates the WE from the sample, providing selectivity by allowing specific ions or molecules to reach the electrode surface while blocking others. It can also protect the electrode from fouling. Common materials are Nafion, cellulose acetate, or PVA.

An electrochemical sensor can work based on different detection methods, as presented in Table 4.8. Each detection method has unique advantages, depending on the target analyte, required sensitivity, and application context.

4.7.4 Optical Sensing

Printed flexible optical gas sensors combine the benefits of optical sensing with the flexibility and scalability of printed electronics. Gas or chemicals sensors can be based on optical sensing techniques (Sect. 4.6). These sensors detect changes in optical properties such as absorption or fluorescence in response to gas/chemicals exposure. They operate based on various optical principles such as absorption spectroscopy, fluorescence spectroscopy, surface plasmon resonance, and others as compared in Table 4.6. Light sources used are typically organic LED (OLED), which are flexible, printable light sources for excitation, or printed quantum dots used for their tunable emission wavelengths and flexibility. As gas-sensing materials can be used: dye-infused polymers, such as transparent polymers, PS or PMMA, doped with gas-sensitive dyes (fluorescent dyes that interact with specific gases), nanomaterials, such as GO, functionalized CNT offering high sensitivity and specificity to certain gases; metal nanoparticles (e.g., Au or Ag nanoparticles used in SPR sensors for their excellent plasmonic properties).

Table 4.8 Main detection methods in electrochemical sensors

Detection method	Working principle	Examples
Amperometric sensing	Measure the current produced by the redox reaction of an analyte at the WE while maintaining a constant potential	The analyte undergoes oxidation or reduction reaction, generating a current proportional to its concentration, e.g., glucose sensors, gas sensor
Potentiometric sensing	Measure the potential difference between the WE and RE without drawing current	Potential difference related to the concentration of the analyte (Nernst equation); pH meters, ion-selective sensors (Na^+, K^+, Ca^{2+}), CO_2, ammonia
Conductometric sensing	Measure changes in electrical conductivity of the solution, which varies based on the concentration of ions present	Used in biosensors, food safety applications, environmental monitoring (e.g., water quality, ions presence)
Impedimetric sensing	Measures the impedance of the system by electrochemical Impedance spectroscopy, EIS, which is affected by surface reactions, mass transfer, and electrode properties	Biosensors for detecting biomolecules like proteins, DNA, and pathogens; corrosion monitoring, battery diagnostics
Voltammetric sensing	Measure the current response as a function of the applied potential Cyclic voltammetry (CV), differential pulse voltammetry (DPV), square wave voltammetry (SWV)	Detection metal ions, organic compounds, and biological molecules (e.g., DNA, proteins)
Coulometric sensing	Measure the total charge passed in a reaction, which is directly related to the quantity of the analyte through Faraday's law (controlled potential or controlled current coulometries)	Trace analysis of metals
Capacitive sensing	Measure changes in capacitance at the electrode interface when the analyte binds or interacts with the electrode surface. The binding event changes the dielectric properties or surface charge	Biosensors, especially for detecting proteins, antibodies, and DNA

4.7.5 *Key Parameters in Printed Flexible Gas and Chemical Sensors*

The main performance parameters of printed gas and chemical sensors are:

Sensitivity—refers to the sensor's ability to detect low concentrations of gas or chemicals. High sensitivity is critical for applications requiring early detection (e.g., hazardous gas detection).

Cross-sensitivity—is the sensor's tendency to respond to gases/chemicals other than the target substance. Low cross-sensitivity ensures that the sensor's response is specific to the target, reducing the occurrence of false positives.

Limit of detection (LOD)—is the lowest concentration of gas or chemical that the sensor can reliably detect. Lower LOD are preferred for applications requiring detection of trace amounts.

Detection range—specifies the minimum and maximum concentrations of gas or chemicals the sensor can detect accurately. A wide detection range allows for flexibility in various applications with different concentration levels.

Selectivity—is the sensor's ability to distinguish a target gas or chemical from other compounds. High selectivity is essential to avoid interference from similar substances.

Response time—is the time the sensor takes to detect a change in gas/chemical concentration. Faster response times are required for applications requiring real-time monitoring.

Recovery time—is the time it takes for the sensor to return to its baseline after exposure to the target gas/chemical. A short recovery time is beneficial for applications that require frequent or continuous monitoring.

Repeatability and reproducibility—repeatability is the sensor's ability to give consistent readings under the same conditions; reproducibility refers to performance consistency across different sensors. Both are essential for reliable data or repeated tests.

Stability and drift—stability measures how well the sensor maintains its performance over time; drift refers to changes in output under constant concentration conditions. Low drift and high stability are necessary for accurate long-term monitoring.

Operating temperature and humidity range—indicate the environmental conditions within which the sensor can perform reliably. Ensuring stability under a wide range of conditions is critical for outdoor and industrial applications.

Response to environmental interference—environmental interference from factors like temperature, humidity, and pressure can affect sensor performance. Robust materials or compensation mechanisms help maintain accuracy in variable conditions.

Hysteresis—refers to the difference in sensor output when exposed to increasing and decreasing concentrations. Low hysteresis is desirable for accuracy, especially in applications with variable concentrations.

Flexibility and mechanical durability—flexibility is critical for integrating sensors into wearable devices or curved surfaces; mechanical durability ensures they can withstand bending and stretching without affecting performance.

Power Consumption—low-power consumption is essential for battery-powered or portable applications, as it extends operational time and reduces the need for frequent recharging.

Form factor—The physical dimensions of the sensor, including thickness, influence its application and integration. Thin, small sensors are ideal for wearable or compact devices.

Cost and scalability—are important for commercial applications, particularly for high-volume production or disposable sensors.

4.7.6 *Comparing Gas and Chemical Sensing Approaches*

Table 4.9 compares the main sensing approaches of flexible printed gas and chemical sensors.

Table 4.9 Comparison of gas and chemical sensing approaches

Sensing Mechanism	Principle	Materials	Advantages	Typical applications
Chemiresistive	Change in electrical resistance due to gas adsorption	Conductive polymers (PANI, PPy, PEDOT), C-based materials (GO, rGO, CNT), metal oxides (SnO_2, ZnO, TiO2, Fe_3O_2), metal nanoparticles, organic–inorganic, I/O, hybrids (perovskites, MOF, zeolites)	High sensitivity, simple design, cost-effective	Environmental monitoring, wearable devices, industrial safety, smart homes
Capacitive	Changes in capacitance caused by the presence of a gas/ chemical	Polymers (PI, PANI, hydrogels), C allotropes, MOF, ionic liquids	High sensitivity, fast response, low-power consumption, simple readout circuitry	Environmental monitoring, industrial control, medical devices,
Electrochemical	Electrochemical reaction between gas and electrolyte, generating a current or voltage	WE: Au, Pt, C allotropes RE: Ag/AgCl CE: Pt, C Electrolyte: PBS, KCl, H_2SO_4; proton-conducting polymers (PBI, SPS, Nafion®)	High sensitivity, specificity, fast response, durable, low consumption, easier to calibrate	Environmental monitoring, wearable devices, HVAC systems, industrial processes
Optical	Interaction of gas with light (absorption, fluorescence, SPR)	Dye-infused polymers (PMMA, PS with fluorescent dyes), nanomaterials (GO, CNT), metal nanoparticles (Au, Ag)	High sensitivity and selectivity, very fast response, non-invasive	Environmental monitoring, industrial safety, medical diagnostics, smart packaging

4.8 Biosensors

Printed flexible biosensors detect and measure the amount of biological analytes, such as enzymes, antibodies, and other biomolecules. The flexibility of these sensors makes them suitable for a wide range of applications, including medical diagnostics, food safety, environmental monitoring, and wearable devices. In these types of sensors, the interaction between the sensor biorecognition element and the target analyte produces a measurable signal. Different types of bioreceptors are used:

(a) enzyme sensors use enzymatic reactions to detect specific substrates (e.g., glucose oxidase for glucose sensing);
(b) immunosensors use antibody-antigen interactions for detection (e.g., proteins and pathogens);
(c) DNA probes detect specific nucleic acid sequences (DNA or RNA) through hybridization; and
(d) aptamers, synthetic oligonucleotides or peptides that bind specific targets, used for detecting various analytes.

The immobilization of bioreceptors uses functionalization techniques such as physical adsorption, covalent bonding, or entrapment in polymers to attach bioreceptors to the electrode surface. Functional layers are used, such as polyethylene glycol (PEG) to reduce non-specific binding and improve sensor specificity, or self-assembled monolayers (SAM) to immobilize bioreceptors on the sensor surface. Flexible printed biosensors operate based on various transduction mechanisms that convert a biological interaction into a measurable signal. Common transduction mechanisms include electrochemical, optical, and piezoelectrical sensing.

4.8.1 *Electrochemical Sensing*

Electrochemical sensing measures changes in current, potential, or impedance due to biochemical reactions, thus detecting biological analytes. These sensors are known for their high sensitivity, specificity, and relatively simple and cost-effective fabrication. An analyte binding to a bioreceptor (e.g., enzyme, antibody) at the electrode surface results in an electrochemical change that is measured. The main types of electrochemical detection methods used in biosensors are amperometric, potentiometric, conductometric, and impedimetric (Table 4.8). The design of electrochemical sensors uses different types of electrodes (Fig. 4.12). The WE are often made of Au, Pt, C, or conductive polymers, which can be functionalized with bioreceptors for specific analyte detection. Electrolytes employed are buffer solutions providing the necessary ionic environment for electrochemical reactions or redox mediators (compounds like ferrocene or ferrocyanide that facilitate electron transfer between the analyte and the electrode).

Electrochemical biosensors show high sensitivity and specificity, being capable of detecting low concentrations of analytes with high accuracy due to specific bioreceptor-analyte interactions and in real time. They have relatively simple and inexpensive fabrication methods, especially when using printed electronics techniques. But long-term stability of bioreceptors can be a concern, with potential degradation over time. Potential interference from other substances in the sample requires careful calibration and compensation. Sensitivity to temperature, pH, and ionic strength of the sample can affect sensor performance.

4.8.2 *Optical Sensing*

Optical biosensors are a powerful method for detecting and measuring biological analytes with high sensitivity and specificity. Optical sensing for flexible printed biosensors combines the advantages of optical detection methods with the flexibility, scalability, and cost-effectiveness of printed electronics. Optical biosensors use light (e.g., absorption, fluorescence, surface plasmon resonance) to detect biological interactions and convert them into measurable signals. An analyte binding to the bioreceptor affects the sensor optical properties, which are detected and measured. Table 4.6 shows the main sensing mechanism under which optical biosensors can operate.

Flexible printed biosensors use OLED or quantum dots as flexible, printable light sources for excitation. Light detectors include printed photodiodes, photomultipliers (e.g., for fluorescence spectroscopy) or CCD/CMOS cameras. The structure of an optical biosensor also includes optical components, such as lenses and mirrors to focus and direct light through the biosensor, optical fibers to transmit light over distances, prisms and gratings, e.g., used in SPR and other spectral analysis techniques. Main sensing materials are dye-infused polymers, and nanomaterials.

Flexible printed biosensors are lightweight, conformable, and portable, making them suitable for a wide range of applications, including wearable health monitors, environmental sensing, point-of-care diagnostics, and food safety. Their ability to provide rapid, accurate, and non-invasive detection makes them a valuable tool in many fields, despite challenges related to durability, complexity, and interference.

4.8.3 *Piezoelectric Sensing*

Piezoelectric sensing in flexible printed biosensors leverages the piezoelectric effect, in which a piezoelectric material generates an electric charge in response to an applied mechanical force (e.g., mass, viscosity), detecting biological interactions. It measures changes in mechanical properties that affect the frequency of a piezoelectric crystal. Analyte binding to the bioreceptor induces a change in mass or viscosity, altering the resonance frequency of the piezoelectric material. These sensors are particularly useful in applications that require high sensitivity, flexibility, and integration into wearable and portable devices.

The piezoelectric effect can be used to detect biological interactions by converting mechanical changes into electrical signals, adopting two main approaches:

- **Direct piezoelectric effect**: generates an electric charge when a force is applied to the piezoelectric material. Binding biomolecules to the sensor surface induces a mechanical stress that generates an electric charge proportional to the mass of the bound biomolecules.
- **Quartz Crystal Microbalance (QCM)**: measures changes in the resonance frequency of a quartz crystal due to mass loading. The binding of biomolecules to the crystal surface changes its mass, altering the resonance frequency, which is measured as an electrical signal.

Piezoelectric materials used are PVdF (high piezoelectric coefficient, flexibility, and durability); ZnO nanostructures (high piezoelectricity, biocompatibility, and ease of integration); PZT (high piezoelectric constant and sensitivity). Flexible piezoelectric thin films are commonly used for integration into wearable devices. Also, nanowire arrays enhance surface area and sensitivity for detecting low concentrations of analytes. Surface functionalization includes immobilization of bioreceptors by techniques such as covalent bonding, physical adsorption, or entrapment in polymers to attach bioreceptors to the sensor surface.

Piezoelectric biosensors have a high sensitivity, being capable of detecting low concentrations of analytes due to significant changes in their mechanical response. They are flexible, can conform to complex surfaces, and maintain performance under bending and stretching, making them ideal for wearable and portable applications. They have a rapid response time, providing real-time monitoring and quick detection of biological interactions. They are also cost-effective. The label-free detection eliminates the need for labeling the analyte, simplifying the assay and reducing costs. Main limitations are their durability, being difficult to ensure long-term stability and resistance to mechanical stress, environmental factors, and repeated use. They also require more complex design and fabrication processes compared to traditional rigid biosensors. Their sensitivity to changes in temperature, pH, and ionic strength, which can affect sensor performance. Depending on the signal processing requirements, power consumption can be a concern.

4.8.4 Key Parameters in Printed Flexible Biosensors

The main performance parameters of printed flexible biosensors are:

Sensitivity—is the sensor's ability to detect low concentrations of the target analyte. High sensitivity is essential for applications where early detection of biomolecules is critical.

Selectivity—refers to the sensor's ability to discriminate the target biomolecule from other similar compounds. High selectivity is critical to ensure accurate readings, avoiding interference from non-target substances, especially in complex biological samples.

Limit of detection (LOD)—is the minimum concentration of an analyte that the sensor can detect reliably. Low LOD is important in detecting trace amounts of biomarkers for early diagnosis and health monitoring.

Dynamic range—defines the concentration span over which the sensor can accurately measure the target analyte. A wide dynamic range is beneficial for applications that may encounter both low and high analyte concentrations.

Response time—is the time required for the sensor to register a change in the analyte concentration. Faster response times are critical in rapid concentration variations, for real-time monitoring applications.

Stability and shelf life—stability refers to the sensor's ability to maintain performance over time; shelf life indicates how long it can be stored without losing functionality. High stability and long shelf life are essential for practical deployment.

Repeatability and reproducibility—Repeatability refers to the sensor's ability to provide consistent readings under the same conditions; reproducibility is the consistency of performance across different sensors. Both are important for reliable data in diagnostic applications.

Biocompatibility—is essential for sensors intended for direct contact with biological tissues or fluids. The materials shall not induce toxicity or immune responses, which is especially important in wearable or implantable biosensors.

Flexibility and mechanical durability—allow the sensor to conform to curved surfaces, such as skin or wearable devices, and withstand bending and stretching. This is important for applications in wearable health monitoring where the sensor may undergo frequent movement.

Hysteresis—is the difference in the sensor's response during increasing and decreasing concentrations of the target analyte. Low hysteresis is desirable for accurate and consistent readings, especially when analyte concentrations fluctuate.

Temperature and pH stability—since biological reactions can be sensitive to temperature and pH changes, the sensor's stability in varying temperature and pH conditions is essential. This stability is particularly important for applications involving bodily fluids or fieldwork in different environments.

Power consumption—Low-power consumption is critical for portable or wearable applications where the sensor may be powered by small batteries. Lower power requirements extend the operational time of the device.

Real-time detection capability—for applications in continuous monitoring, real-time detection capability is essential. This parameter is critical for wearable biosensors that provide continuous health data or environmental monitoring.

Regeneration and reusability—Some applications benefit from reusable biosensors that can be regenerated for multiple uses. Reusability can reduce costs, though disposable sensors are preferred for some applications, like diagnostics, to avoid cross-contamination.

Form factor—The physical dimensions of the sensor, including its thickness and shape, influence how it integrates into devices. Compact, thin sensors are preferred for wearables, implantables, and applications where space is limited.

Table 4.10 Comparison of biosensing approaches (WE—working electrode, RE—reference electrode)

Sensing principle	Transduction mechanisms	Materials	Advantages
Electrochemical: measures changes in current, potential, or impedance due to biochemical reactions	Amperometry, potentiometry, conductometry, impedimetry	WE: Au, Pt, C, conductive polymers. RE: Ag/AgC Bioreceptors; enzymes, antibodies, DNA probes, aptamers. electrolytes: buffer solutions, redox mediators	High sensitivity and specificity, rapid response time, miniaturization and portability, cost-effective
Optical: measures changes in light properties (absorption, fluorescence, SPR) due to biochemical interactions	Absorption spectroscopy, fluorescence spectroscopy, surface Plasmon resonance (SPR), Optical waveguide	Light Sources: OLED, photodiodes, photomultipliers, CCD/ CMOS cameras. sensing materials: fluorescent dyes, metal films (gold, silver), optical waveguides	High sensitivity and selectivity, rapid response time, non-invasive, label-free detection (SPR), multiplexing capability
Piezoelectric: measures changes in mechanical response or frequency due to mass loading from biomolecule binding	Direct piezoelectric effect, quartz crystal microbalance (QCM)	Materials: PVdF, ZnO nanostructures, PZT. Bioreceptors: enzymes, antibodies, DNA probes, aptamers	High sensitivity, flexibility, and integration into wearable devices, rapid response time, label-free detection

Cost and scalability—For commercial viability, low production costs and scalability are essential, especially for disposable biosensors. Printed electronics techniques often make biosensors more affordable and suitable for high-volume production.

4.8.5 Comparing Biosensing Approaches

Table 4.10 compares the main sensing approaches of flexible printed biosensors.

4.9 Organic Field-Effect Transistor, OFET, Sensors

An organic field-effect transistor (OFET) is a type of transistor that uses organic semiconductors instead of traditional inorganic materials (e.g., silicon). These sensors use organic materials (like polymers and small molecules) as semiconductors, which are

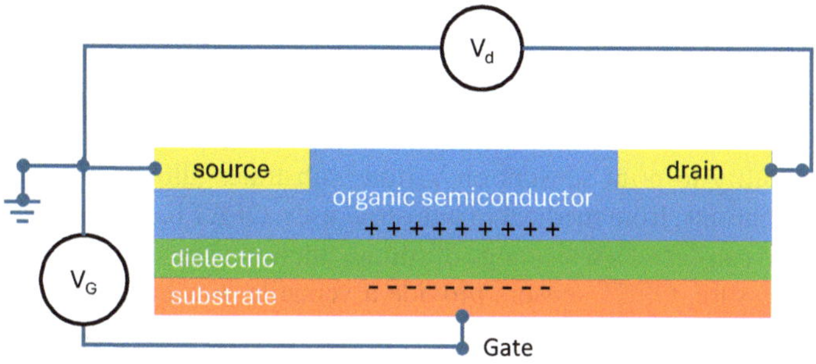

Fig. 4.13 Typical OFET design

compatible with flexible substrates. Organic semiconductors (OS) are easy to process, can be dissolved in solvents, and are compatible with printing techniques.

The basic structure of an OFET consists of three main components: the source (S), the drain (D), and the gate (G) (Fig. 4.13). The S is where the charge carriers (electrons or holes) enter the device; the D is where the charge carriers exit the device; G is a control electrode that modulates the flow of charge carriers between the S and D. OS is the active material that conducts charge carriers. The dielectric layer is a thin insulating layer between G and OS, which allows gate voltage control without direct current flow.

When voltage is applied, G controls the movement of electrons (or holes) from S to D through the OS layer. If V_G induces enough charge in the channel, the S-D current flows (S-D current). By applying a voltage at the gate terminal between the S and D electrodes (V_d), an electric field is generated across the gate dielectric, which induces a charge accumulation or depletion in the OS layer near the semiconductor/dielectric interface. The amount of current depends on several factors, including the gate voltage, the thickness of the organic semiconductor, the quality of the gate dielectric, and the mobility of the charge carriers in the organic material.

When no gate voltage is applied ($V_G = 0$), the current between the S and D is minimal or zero (off-state). When a sufficient gate voltage is applied (positive for n-type or negative for p-type semiconductors), a conducting channel is formed between S and D, allowing current to flow (on-state). The transistor can act as a switch or amplifier by modulating the gate voltage to control the flow of current between S and D. The gate voltage at which the OFET begins to turn on and conduct significant current is called the threshold voltage (V_{th}). Below V_{th}, the current remains minimal. In some cases, the properties of the OS or external factors, like temperature or environmental exposure, can shift the threshold voltage, affecting the transistor's performance.

In general, OFET operates based on changes in electrical characteristics (current or voltage) caused by interactions with the variable to be detected. This mechanism allows the transistor to switch electrical signals on and off (as in a traditional FET). OFET generally has lower charge carrier mobility than their inorganic counterparts, but recent advancements are improving their performance. Environmental factors like humidity and oxygen can degrade the OS, though encapsulation techniques are being developed to enhance durability.

OFET sensors offer several advantages, including flexibility, lightweight design, and the ability to be produced at lower costs using techniques like printing on large areas. The ability to print sensors on various substrates allows customization for specific applications, from medical to industrial uses. OFET-based sensors have high sensitivity and can detect a wide range of stimuli, including gases, chemicals, temperature, and pressure. OFET sensors provide a versatile, low-cost, and flexible platform for sensing technologies for wearable health monitoring (e.g., tactile sensing, heart rate, sweat composition, or hydration levels), environmental monitoring (e.g., pollutants, toxic gases, pH, humidity), chemical and biosensors (e.g., VOC, glucose, lactate, pathogens, biomarkers, DNA/RNA, proteins).

4.9.1 Types OFET Sensors

OFET sensors can be classified based on the type of materials used, their structure, and the stimulus detected.

- **Materials-Based Classification**

Based on the OS materials, OFET is divided into (Fig. 4.14):

- p-Type OFET (Fig. 4.14a)—uses OS that primarily conduct holes (positive charge carriers), e.g., pentacene, polythiophenes; they have higher performance and stability than n-type OFET.
- n-Type OFET (Fig. 4.14b)—uses OS that conduct electrons (negative charge carriers), like fullerenes, perylene diimides; they are more challenging to develop because OS often exhibit lower electron mobilities.
- Ambipolar OFET (Fig. 4.14c)—conducts both electrons and holes, depending on the gate voltage; they combine the properties of both n-type and p-type OFET, making them versatile in applications that require dual charge conduction.

- **Structure-based classification**

OFET can be designed using different structures, each with specific configurations of electrodes and layers that influence their performance. The main types of OFET structures are shown in Fig. 4.15.

In **BGBC—Bottom-Gate, Bottom-Contact** structure (Fig. 4.15a), both G and S/D electrodes are located beneath the OS layer. The gate electrode is first deposited on the substrate, followed by the dielectric layer, then S and D electrodes, and finally the OS on top. This structure allows for better control over the thickness of the OS layer. It also provides a smooth interface between the OS and the S/D contacts, leading to improved charge injection. The overlap between the OS and the electrodes can sometimes result in higher contact resistance, reducing overall device performance.

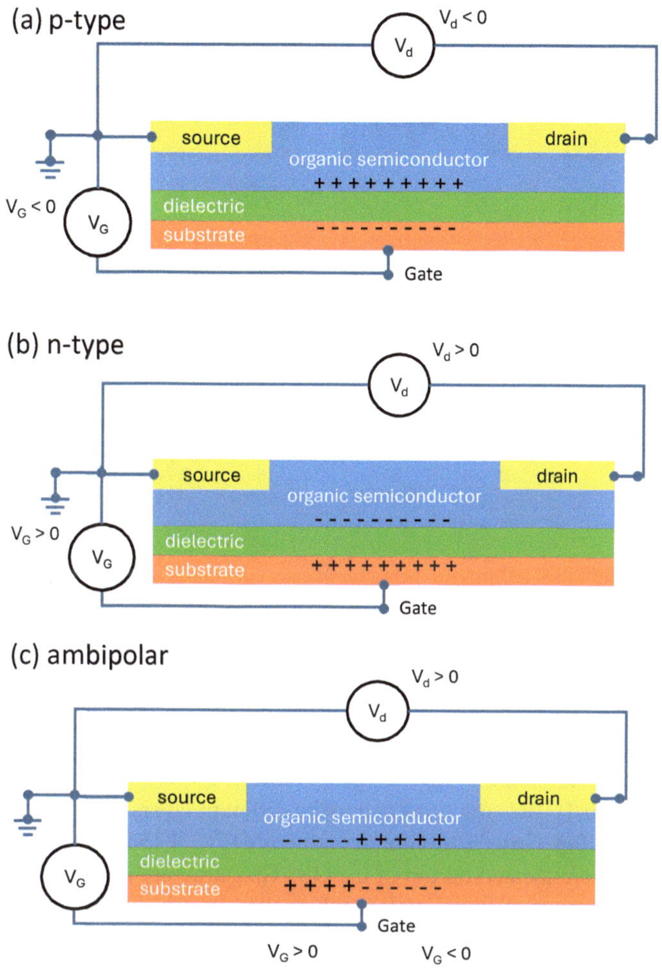

Fig. 4.14 Classification of OFET based on the type of OS material

In **BGTC—Bottom-Gate, Top-Contact** structure (Fig. 4.15b), G electrode is still located at the bottom of the device, but S and D electrodes are placed on top of the OS. The dielectric layer is deposited over G, and then OS and top electrodes are added. BGTC structures often result in better electrical performance compared to BGBC, as the top contacts typically allow for lower contact resistance between OS and the electrodes.

In **TGBC—Top-Gate, Bottom-Contact** structure (Fig. 4.15c), S and D electrodes are located beneath the OS, and the G electrode is placed above the OS layer. G is separated from the OS by a dielectric layer. TGBC structures are well-suited for flexible electronics, as G electrode can be easily integrated into flexible substrates.

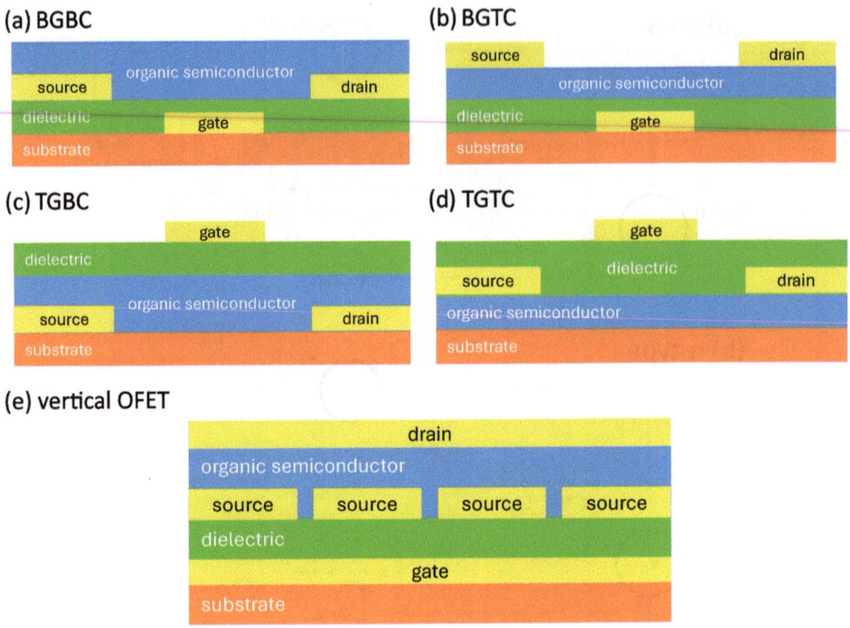

Fig. 4.15 Types of OFET structures

This structure provides better stability and protection of the OS layer from environmental factors like humidity and oxygen. Achieving low contact resistance can be more difficult in TGBC structures due to the different layers' alignment.

In **TGTC—Top-Gate, Top-Contact** structure (Fig. 4.15d), both G electrode and S/D electrodes are located on top of the OS. This structure is less common but has been explored for specific applications where the active layer needs to be protected. This structure allows for better integration into multilayer systems and provides protection to the underlying layers from exposure to air. TGTC structures can be advantageous in flexible or rollable devices and situations where the transistor needs to be encapsulated.

In **vertical OFET** (Fig. 4.15e), the charge transport occurs vertically between electrodes, rather than laterally as in conventional OFET. This allows for shorter channel lengths, leading to potentially faster switching speeds and higher current densities. Nevertheless, the interfaces between OS and the electrodes can have high contact resistance, which affects sensor performance. Also, this vertical design adds complexity to the fabrication process, especially when scaling for mass production. The challenge is to precisely control the thickness of the organic layers and the alignment of electrodes to ensure efficient charge transport.

BGBC and TGBC structures are co-planar structures; i.e., S and D electrodes are positioned in the same plane relative to the OS. BGTC and TGTC are staggered structures, where S and D electrodes are placed on opposite sides of the OS layer.

These configurations affect the charge injection, transport, and overall performance of the device. Table 4.11 compares these types of OFET configurations.

Unlike traditional planar OFET, vertical OFET (Fig. 4.15e) uses a vertical channel for charge transport between S and D, enabling more compact devices and higher current densities. This structure is beneficial for applications requiring high current outputs in small devices, such as organic displays or circuits. Vertical OFETs are being explored for organic light-emitting devices (OLED) and flexible electronic applications.

- **Application-based classification**

Depending on the application and the nature of the analytes they are designed to detect, OFET sensors can be classified as shown in Table 4.12.

Each type of OFET sensor has unique design requirements based on the target analyte, substrate material, and environmental conditions, with ongoing research to improve their sensitivity, stability, and scalability for broader applications.

Table 4.11 Comparison of OFET configurations

OFET structure → main feature	Co-planar	Staggered
Electrode positioning	S and D electrodes positioned in the same plane on the same side of the OS. Both the OS and the S/D electrodes are deposited on the dielectric layer and substrate, ensuring that the charge carriers do not have to pass through multiple layers during injection	S and D electrodes are placed on opposite sides of the OS layer. One set of electrodes is positioned below OS layer, while the other set above it. Charge carriers injected from S have to traverse through the OS layer before reaching the D, making the charge transport path longer
Charge injection	Charge injection occurs directly at the interface between S/D electrodes and OS; lower contact resistance; better charge injection efficiency, particularly in top-contact OFET, where the electrodes are deposited after the OS layer	Charges are injected into the OS from one side and must travel through it before reaching D electrode; higher resistance; reduced efficiency in charge injection; often seen in bottom-contact OFET, where the OS is deposited after the electrodes
Device performance	Provide better performance for high-mobility OFET, due to the reduced resistance at the interface between the OS and electrodes; lower contact resistance, which improves current flow and switching speed; high-performance applications	Higher resistance due to the longer charge transport path; slower device switching; Higher threshold voltages; more stable, with better control over the alignment of the OS with the electrodes
Application suitability	Applications requiring high speed and low contact resistance, such as displays and sensors where performance is critical	Applications where stability and manufacturing simplicity are prioritized, or where precise control over layer deposition is necessary

Table 4.12 Main types of OFET sensors

Type OFET	Description	Application
Chemicals and gases sensors	OS layer is functionalized to react with specific gases, altering the transistor's current	Detection of VOC, toxic gases (NH_3, NO_2, CO), O_2
Ion sensors Ion-sensitive field-effect transistors (ISFET)	When ion concentration changes, the current through the transistor will change accordingly. The solution may be used as the G electrode	Biological, chemical, environmental analysis; monitoring pH or specific ionic concentrations (e.g., K, Ca, Na)
Biosensors	A biorecognition layer (e.g., antibody, enzyme) is immobilized on the surface of the OS, interacting with the target analyte and changing the electrical properties of the OS layer; the change in current or voltage between S and D is correlated with the concentration of the target analyte	Highly sensitive to proteins, bacteria, DNA/RNA, enzymes (e.g., glucose), antigens, or pathogens
Pressure, force and strain sensors	The applied force causes physical changes in the OS or the layers that make up the transistor, which affect the electrical properties of the OFET, such as charge carrier mobility, dielectric modulation, and contact resistance between S/D and OS	Wearable devices; motion tracking, tactile sensors (e.g., artificial skin, soft robotic prosthetics), health monitoring (e.g., vital signals)
Temperature sensors	Respond to changes in temperature by modifying the current flow through the transistor, i.e., by monitoring variations in the charge mobility of the OS material	Environmental monitoring, wearable devices, thermal sensing
Optoelectronic sensors (light/optical) Photodetectors (VIS, UV, IR light); OLET (organic light-emitting transistor)	Uses OS that respond to optical stimuli, detecting changes in light intensity or wavelengths; they operate by converting light into an electrical signal (photon absorption, charge separation, current modulation), leading to a measurable change in the current between S and D electrodes	Imaging systems, environmental monitoring, photodetectors, light sensors, flexible displays, electronic paper, optical sensor, medical diagnostics
Humidity sensors	OS electrical conductivity or the dielectric constant of the dielectric layer change with the humidity level; water molecules interact with the OS or the dielectric layer in the device, leading to changes in the current flowing between S-D	Agricultural applications, environmental control systems, smart home devices

(continued)

Table 4.12 (continued)

Type OFET	Description	Application
Multifunctional OFET sensors	Detect multiple parameters simultaneously, such as pressure, humidity, and temperature; Integrate sensing and actuation capabilities (e.g., OLET—organic light-emitting transistor); integration of different sensing layers or materials into a single flexible OFET device	Complex monitoring applications, environmental sensing systems, wearable electronics for health, smart skins
Hybrid OFET sensors	Combine the best features of organic and inorganic materials, offering improved performance, stability, and versatility	Flexible electronics, wearable devices, optoelectronics, energy harvesting
Field-effect capacitive sensors	Measure changes in capacitance between a sensing electrode and G, modulating the OFET electrical properties, e.g., V_G; G electrode acts as one of the plates of a capacitor, with the external environment influencing the other side of the dielectric layer; the external stimulus causes a change in the dielectric constant of the material between G electrode and the external environment, modulating the capacitance, shifting the gate voltage, leading to a change in the current between S and D	Interactive electronics, touch panels, touch or proximity sensing, pressure sensing, other user interface applications

4.9.2 OFET Sensing Mechanisms

The main sensing mechanisms in OFET-based sensors are primarily based on changes in the electrical properties of the OS material in response to external stimuli, namely by modulating the current between S and D electrodes. The change in current (I_{DS}) or threshold voltage (V_{thr}) forms the basis of detection (see Sect. 4.9.3). OFET sensors employ various sensing mechanisms, such as, charge carrier modulation, surface interaction, dielectric sensing, ion sensing, optoelectronic sensing, and mechanical strain detection, to respond to different external stimuli:

- **Charge carrier modulation** is the fundamental mechanism by which OFET operates. In the OS layer, the movement of charge carriers (electrons or holes) is modulated by applying a voltage at the gate terminal. When the sensor detects an external stimulus (e.g., gas, light, biomolecules), it affects the mobility or concentration of these carriers, altering the current flow between the S and D electrodes.
- In the **surface interaction (adsorption/desorption)** mechanism, the analyte adsorbs onto the OS layer's surface. This interaction alters the material's electronic properties, such as charge carrier mobility, by either donating or withdrawing electrons or holes. Desorption reverses the effect. Chemical and gas sensors primarily use this mechanism.

- The **dielectric sensing** mechanism involves changes in the dielectric properties of the gate dielectric or of the OS-dielectric interface. When the dielectric constant changes due to the presence of an analyte, it influences the capacitance of the device and, in turn, modulates the transistor's performance. OFET-based humidity sensors and biosensors (for detecting proteins or DNA) often rely on this mechanism. The analyte alters the dielectric properties, affecting the gate's ability to control the charge carriers in the OS layer.
- **Ion sensing (electrochemical)** mechanism is typical of electrochemical sensors, where the OFET structure is designed to respond to specific ions. These sensors detect ions by sensing changes in the ionic environment around G or in the electrolyte solution, which alters the current between S and D terminals. This mechanism is used in biosensors for detecting pH changes, ions, or other charged species in a solution, as well as in glucose sensors for medical diagnostics.
- In **optoelectronic sensing mechanism**, the OFET also functions as an optoelectronic sensor where external light modulates the carrier mobility in the OS layer. Photons excite electrons or holes, generating a photocurrent that can be measured. This mechanism is used in photodetectors and light-sensing applications such as OLET and phototransistors.
- In **mechanical (strain/pressure)** sensing, the mechanical deformation (strain or pressure) applied to the OS or to the flexible substrate changes the transistor's electrical properties. This is typically due to changes in the microstructure of the OS or variations in the distance between molecules, affecting charge mobility. OFET-based pressure and strain sensors are used in flexible electronics and wearable devices to detect pressure, bending, or stretching.

4.9.3 OFET Operation Principles

The operational principle of an OFET is based on the modulation of charge carrier flow (electrons or holes) through an OS channel, controlled by an electric field applied via the G electrode. When a voltage is applied between S and D, charge carriers are injected from the source into the OS (charge injection). These carriers then travel through the OS toward the D (charge transport). The mobility of the carriers is influenced by the organic material's molecular structure, purity, and any defects within the OS layer. Carriers are extracted at D (charge extraction), and the flow of these charges constitutes the current I_{DS}. The operation of OFET relies on the application of two potentials (Fig. 4.13): one on the D electrode (V_d) and the other on the G electrode (V_G), being generated the current I_{DS}. Figure 4.16 shows a typical plot of I_{DS} as a function of V_d, at different fixed values of V_G (transfer curve).

Two regions of interest can be identified. A **linear regime** in which I_{DS} varies almost linearly with V_d, and a **saturation regime** showing a constant V_d value. When this occurs, the channel is said to be **pinched off**. In fact, the operation of the OFET can be categorized into different regimes depending on the applied V_G and V_d, as highlighted in Fig. 4.16. These operation regimes dictate the behavior of I_{DS} and are

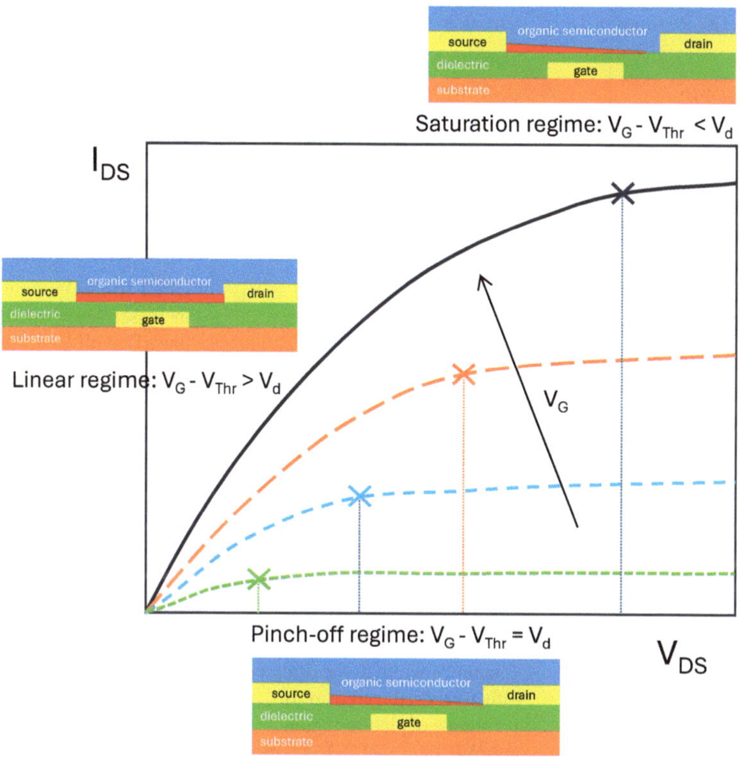

Fig. 4.16 Typical transfer curve ($I_{DS} - V_{DS}$ curves) of an OFET at different V_G

essential for understanding the device's performance in different applications, such as switching and amplification. Table 4.13 shows the main regimes of operation of OFET.

Each of these regimes plays a critical role in various applications, ranging from switching and digital logic to analog amplification and low-power sensor systems.

4.9.4 Design of OFET Sensors

The design of OFET sensors is highly application-dependent, requiring careful consideration of materials, structural components, functionalization and fabrication techniques, and specific application requirements to meet specific sensing demands.

- **Materials**
  - **Substrate Selection**—The substrate supports the OFET sensor and significantly impacts its flexibility, durability, and application. Flexible substrates include PET (flexible, lightweight, transparent), PI (excellent thermal stability

Table 4.13 Regimes of OFET operation

Regime	Condition	Operation
Cutoff (OFF) $V_G \approx 0$ (effective switch)	$V_G < V_{thr}$ Gate-source voltage is below the threshold voltage	OFET in its OFF state: V_G is not high enough to induce charge carriers in OS; no conductive channel formed between *S* and *D*; only a negligible leakage I_{DS} flows
Linear or Ohmic (ON) (analog amplification)	$V_G > V_{thr}$ and $V_d < V_G - V_{thr}$ Gate-source voltage is above threshold, but drain-source voltage is low 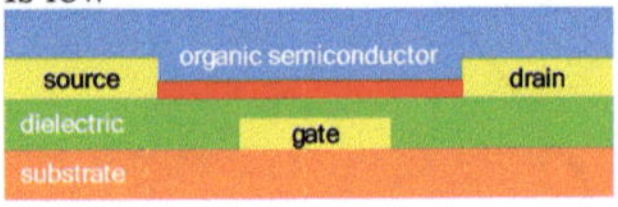 $I_{DS(\text{lin})} = \mu C_i (W/L)(V_G - V_{thr})$ where μ is the carrier mobility, C_i the gate dielectric capacitance, W/L the width-to-length ratio of the *S-D* channel	OFET in its ON state: I_{DS} increases approximately linearly with V_{DS}; channel fully formed, and charge carriers move through it due to the applied V_d; the resultant I_{DS} depends on the applied V_G and the intrinsic properties of the OS
Pinch-Off	$V_d = V_G - V_{thr}$ Channel's width at the drain end is effectively reduced to zero, forming a depletion region 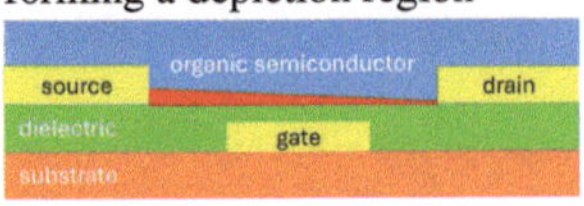	The voltage at G is high enough to induce a conducting channel between *S* and *D*, but as V_{DS} continues to increase, the electric field in the channel becomes stronger near the drain. Eventually, the channel near the drain becomes "pinched off," meaning it no longer allows the charge carriers to pass through easily, and the current stops increasing linearly with V_{DS}

(continued)

Table 4.13 (continued)

Regime	Condition	Operation
Saturation (active) or threshold voltage (amplification)	$V_G > V_{thr}$ and $V_d \geq V_G - V_{thr}$ Gate-source voltage is above threshold, but drain-source voltage is high 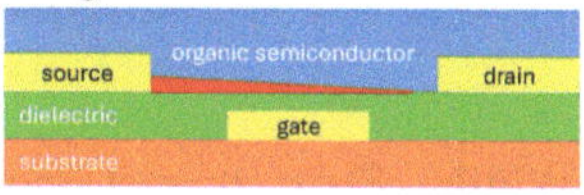 $I_{DS(sat)} = \mu C_i(W/L)(V_G - V_{thr})^2$	OFET enters the saturation region (V_{thr}), where I_{DS} becomes less dependent on V_d and more controlled by the gate voltage V_G. The channel near the drain "pinches off," meaning it narrows and eventually becomes depleted of charge carriers, resulting in a constant current. V_{thr} value depends on the material properties and the device design
Subthreshold region (energy efficient; energy-harvesting, ultra-low-power circuits)	$V_G < V_{thr}$, but a small current flows (part of the cutoff regime)	In this region a small leakage current can still flow even when the gate voltage is slightly below V_{thr}. The current in this region follows an exponential relationship with V_G, making this region important for low-power applications

and flexibility), paper (eco-friendly, cost-effective), textiles. The substrate's mechanical properties are crucial, especially for applications where the device is exposed to bending, stretching, or wear (e.g., wearables, robotics).

- **Organic Semiconductor layer**—is the active material that controls the charge transport in the transistor and is central to its sensing capabilities. Main types of organic semiconductors include small molecules (e.g., pentacene, rubrene, which offer high charge mobility but are less flexible) and conjugated polymers (e.g., poly(3-hexylthiophene)—P3HT). Materials with higher mobility lead to better performance, but it must be balanced with flexibility and processability. OS are often sensitive to environmental factors (e.g., oxygen, humidity), so encapsulation or protective coatings are often needed to improve long-term stability.
- **Dielectric Layer**—insulates the gate electrode from the semiconductor layer and plays a key role in the gate modulation of the transistor. Main materials used are inorganic dielectrics (e.g., SiO_2, Al_2O_3, for high-performance but rigid sensors), organic dielectrics (e.g., PMMA, PVA, offer flexibility but may have low dielectric constants), high-k dielectrics (e.g., hafnium oxide, HfO_2, are used to reduce operating voltages). The dielectric layer thickness impacts the sensor's threshold voltage and operational stability. Thinner dielectrics reduce the operating voltage, which is desirable for wearable and portable devices.

- **Electrodes (S, D, G)**—The electrodes form the electrical contacts of the OFET and determine the injection of charges into the OS layer. Several materials are used, such as Au, Ag (frequently used for S and D electrodes due to their high conductivity), conductive polymers (like PEDOT: PSS are flexible, and printable but less conductive), carbon-based electrodes (e.g., graphene are emerging options for low-cost, flexible, and transparent electrodes).

- **Structural Components**
  - **Channel Length and Width**—Short channel lengths enhance performance by reducing the charge transport time, but they are more difficult to manufacture using printing techniques. The channel width can be adjusted to influence current flow for different sensor applications.
  - **Electrode Patterning**—Printing techniques like inkjet printing, screen printing, or photolithography can be used to define the electrode pattern.

- **Functionalized Layer**
  - For sensors, especially chemical or biosensors, the active layer often requires functionalization to target specific analytes.
  - Chemical functionalization involves attaching receptor molecules (e.g., antibodies, enzymes, chelating agents) to OS layer to enhance selectivity.
  - Self-Assembled Monolayers (SAM) can be used to control surface properties and enhance sensor sensitivity by promoting better charge transfer between the analyte and the semiconductor.
  - For biosensors, the functionalization must ensure compatibility with biological samples (e.g., blood, saliva) without interfering with the sensor's performance.

- **Fabrication processes**
  - The fabrication process significantly influences its performance, reliability, and operational characteristics of OFET.
  - The method used to **deposit the dielectric layer** (e.g., spin coating, atomic layer deposition) affects its uniformity and thickness. Defects or impurities in the dielectric can lead to higher gate leakage currents or increased threshold voltages, reducing the overall performance of the device.
  - The **deposition method of the electrodes** (e.g., thermal evaporation, printing techniques) affects the contact resistance between the electrodes and OS. Poor contact can lead to higher contact resistance, which reduces the device's current output and performance.
  - The method used to **deposit the OS** (e.g., spin coating, inkjet printing, blade coating) impacts the film morphology and the ordering of the molecules. Poor molecular alignment or non-uniform films can result in lower charge carrier mobility. Optimized deposition techniques, such as solution shearing or blade coating, can improve molecular alignment and crystallinity, leading to better charge transport properties and higher device performance.
  - **Surface treatment of the substrate** can improve the adhesion and crystallinity of the OS, leading to higher device performance and stability.

- **Annealing of the deposited OS layers and electrodes** helps to improve the printed/deposited material crystallinity and to remove residual solvents. This process can significantly increase charge carrier mobility and reduce the threshold voltage. Improper annealing can cause material degradation, while optimized annealing can enhance molecular ordering and reduce defects in the semiconductor layer.
- Since organic materials are typically sensitive to environmental degradation (e.g., oxygen, humidity), **encapsulation** is critical to ensure device stability and long-term performance. Main encapsulation materials include PDMS (flexible and transparent, suitable for stretchable sensors), epoxy or other polymers (for rigid, long-term protection against humidity and oxygen). Proper encapsulation is necessary to protect the sensor, especially for applications in harsh environments or wearables that come into contact with the skin.

OFET sensors typically operate at lower voltages compared to their inorganic counterparts. However, for battery-powered or energy-harvesting devices, minimizing power consumption is essential. Using high-k dielectric materials and optimizing channel length helps reduce operating voltages, making OFET sensors more efficient for portable applications. OFET sensors require circuitry to capture, process, and transmit the sensor data. Integration with low-power, flexible circuits for wireless communication (e.g., Bluetooth, NFC) is critical for wearable devices and IoT applications. Weak signals may require amplification to be usable in practical applications.

4.9.5 Key Parameters in Printed Flexible OFET Sensors

Key performance parameters are tuned during the design process, and the most important for determining the efficiency of the OFET sensors are:

Field-effect mobility—the mobility of charge carriers in the OS determines how quickly the carriers can move through the device, which can be determined from the linear or the saturation regimes of the transfer curve (Fig. 4.16). A large difference between these values may indicate that contact resistance is an issue. Higher mobility results in faster switching speeds and better performance.

Threshold Voltage (V_{thr})—is the minimum voltage for the operation of an OFET, determining when the device turns on. Lower threshold voltages generally lead to more energy-efficient devices. Nevertheless, V_{thr} is not zero due to the presence of impurities and high surface roughness. Lowering V_{thr} improves sensitivity, especially for wearable and low-power applications.

I_{ON}/I_{OFF} ratio—ratio between the currents in ON and OFF states. A high I_{ON}/I_{OFF} ratio is desirable for efficient switching, and better performance of the sensor, minimizing leakage currents. I_{on} value is affected by mobility, type and density of charge carriers, capacitance of the dielectric material, and the operating voltages. I_{off} depends upon the dielectric material, and the gate-contacts configuration.

Response time—Fast response times are important for real-time monitoring applications.

Stability and repeatability—Ensuring consistent sensor performance over multiple cycles and long durations is key in several applications.

4.9.6 Materials for OFET

OFET uses a variety of materials, each chosen for its specific role in the device structure. The choice of materials in OFET is fundamental for tuning device performance, flexibility, and manufacturability. Organic materials offer flexibility, solution process ability, and potential low-cost manufacturing, but improving their electrical performance and stability remains a key challenge in this field. The key materials involved in OFET include OS, dielectric layers, electrodes, and substrates, as shown in Table 4.14.

4.10 Surface Acoustic Waves, SAW, Sensors

In SAW sensors, acoustic waves travel along the surface of a piezoelectric material detecting changes in physical, chemical, or biological conditions of the surface. These sensors are highly sensitive and are used in various applications, including gas sensing, biosensing, pressure, temperature, and humidity sensing. SAW sensors operate based on the interaction of acoustic waves with the sensor's surface and the changes induced by the environment or the target analyte. They consist of a piezoelectric substrate printed with two interdigitated electrodes (IDE), and an active zone between them (Fig. 4.17).

In a SAW sensor, a piezoelectric material (like quartz, lithium niobate, or lithium tantalate) generates surface acoustic waves. An input interdigital transducer (Tx) is used to convert an electrical signal into a surface acoustic wave via the piezoelectric effect. The generated acoustic wave propagates along the surface of the sensor (active zone). When the wave interacts with the surrounding medium (e.g., gas, liquid, or biomolecules), the wave properties, such as velocity, amplitude, and phase, change due to mass loading, elastic deformation, or changes in electrical properties. The output interdigital transducer (Rx) detects the modified surface acoustic wave and converts it back into an electrical signal. The changes in the wave's properties are analyzed to determine the presence or the concentration of the target substance. Main advantages of SAW sensors are:

a. Wireless Operation: SAW sensors can operate wirelessly, making them suitable for harsh or inaccessible environments.
b. High Sensitivity: These sensors are highly sensitive to small changes in mass, pressure, or temperature, making them ideal for precision sensing applications.

Table 4.14 Materials for OFET sensors

Component	Main function	Used materials
Organic semiconductor, OS	Charge transport	Small molecules: Pentacene, Rubrene, Tetracene, Anthracene, Fullerenes C60 (n-type OS)
		Conjugated polymers: Poly(3-hexylthiophene), P3HT (p-type), Poly(2,5-bis(3-hexadecylthiophen-2-yl)thieno[3,2-b]thiophene—PBTTT, Polyfluorenes, n-type polymers (e.g., naphthalene diimide-based polymers)
Dielectric	Insulates G electrode from OS; key role in G capacitance, influencing the switching behavior	Organic dielectrics: Poly(methyl methacrylate)—PMMA, Polyvinylphenol—PVP
		Inorganic dielectrics (better performance but less flexibility): SiO_2, Al_2O_3 Hybrid dielectrics (combinations of I/O materials)—used to balance flexibility and performance
Electrodes (S, D, G)	Used to inject or collect charges in the SC	S, D: Au, Ag, Al, conductive polymers (e.g., PEDOT) G: metal, printed conductive inks, graphene
Substrates	Provides mechanical support for the entire OFET structure	Flexible substrates: PET, PI, PEN, paper, fabrics
Interfacial layers	Improve charge injection and device performance, adding interfacial layers between electrodes and OS; reduce contact resistance	SAM: often applied to modify the surface properties of the electrodes, improving contact and reducing contact resistance Bis(trimethylsilyl)amine, an organosilicon compound known as Hexamethyldisilazane (HMDS), is used to passivate surfaces and reduce the influence of trap states at interfaces

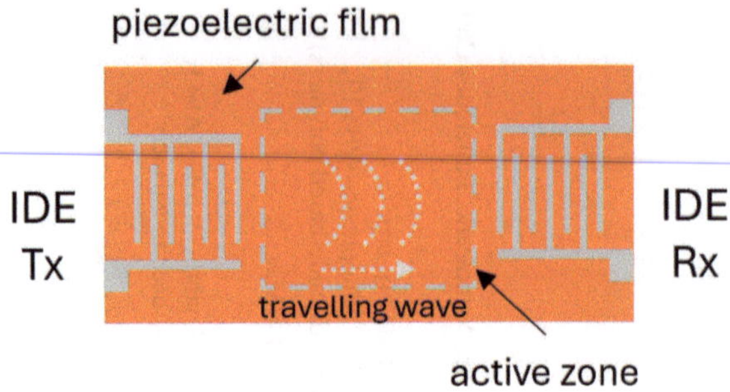

Fig. 4.17 Surface acoustic wave sensor

c. Passive Sensing: SAW sensors do not require an external power source, as they are passive devices that convert energy between electrical and acoustic forms.
d. Robustness: SAW sensors are robust and can operate in a wide range of environmental conditions, including high temperatures and corrosive environments.

SAW sensors are versatile and offer a reliable, low-power option for sensing in a wide range of applications. Their passive and wireless capabilities make them increasingly popular in emerging fields like Internet of Things (IoT) and remote sensing applications. Main issue with current SAW sensors in flexible electronics is the brittleness of the piezoelectric material, typically a ceramic thin layer of quartz, lithium niobate, or lithium tantalate. Also, flexible SAW devices may exhibit significant frequency shifts and signal interferences when deformed (mainly due to changes in the acoustic wave velocity and deformation of the IDE).

4.10.1 Types of SAW Sensors

SAW sensors can detect and measure various physical variables. Based on their application and the physical parameters they measure, they can be classified into various types, as shown in Table 4.15.

The main types of SAW sensors can be categorized based on their configurations and the physical properties they measure. The structure of a SAW sensor determines the type of acoustic wave generated and how it interacts with the surrounding environment or the medium under investigation. Main types of SAW sensors based on their structure and wave propagation characteristics are presented in Table 4.16.

4.10.2 SAW Sensing Mechanisms

SAW sensors operate by detecting changes in the propagation of acoustic waves along the surface of a piezoelectric material due to variations in physical, chemical, or biological conditions. The sensing mechanisms of SAW sensors are based on how

Table 4.15 Main sensing variables with SAW sensors

Sensing variable	Principle	Example of application
Temperature	Measure temperature changes based on the velocity of surface acoustic waves and thus the frequency of the signal, which varies with temperature	Wireless SAW temperature sensors are used in industrial and automotive applications
Pressure	Variations in pressure change the mechanical stress on the SAW sensor's surface, altering wave velocity, shifting the frequency and phase of the surface acoustic wave and its amplitude	Used in tire pressure monitoring systems (TPMS) in vehicles
Humidity	A hygroscopic layer is applied to the sensor surface; changes in humidity cause variations in the mass or electrical conductivity, altering the characteristics of SAW sensor	Used in environmental monitoring, weather stations, and smart home systems
Gas	Detect specific gases by applying a selective coating on the sensor's surface that interacts with the target gas and causes changes in mass, electrical conductivity, or other properties, altering the wave characteristics	Used in air quality monitoring and industrial safety systems
Chemicals	Detect specific chemicals by applying a specific coating on sensors surface, which interacts with the chemical and causes changes in mass or viscoelastic properties on the sensor's surface	Used in medical diagnostics, environmental monitoring, and chemical process industries
Biosensors	Functionalized with biomolecules (like antibodies, enzymes, or DNA), the sensor detects biological analytes; the binding of these analytes causes changes in the mass and surface conductivity, leading to shifts in the wave's properties	Used in medical diagnostics, such as detecting pathogens, proteins, or DNA sequences
Mass loading	Any added mass to the sensor's surface (e.g., particles, molecules, liquids) changes the wave's propagation velocity and frequency	Applied in biosensing, aerosol detection, and thin-film deposition monitoring
Viscosity and density of liquids	The interaction of the SAW sensor with a liquid changes the damping of the wave, allowing for the measurement of viscosity and density	Used in monitoring fluid properties in chemical processes, fuel quality assessment, and food industry applications
Strain and stress	Mechanical strain or stress alters the substrate's properties, affecting wave velocity or frequency	Used in structural health monitoring, load sensing, and material testing

Table 4.16 Main types of SAW sensors based on their configuration and wave propagation characteristics

Type	Description	Examples of application
Delay line SAW sensors 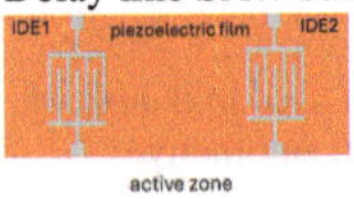	An input transducer generates the acoustic wave and an output transducer receives it after a delay. The delay in the wave propagation changes in response to external stimuli	Gas concentration, humidity, and temperature
Resonator SAW sensors 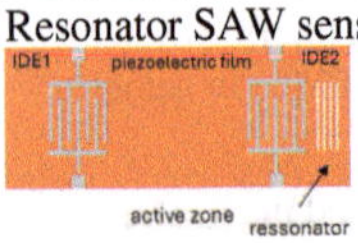	Resonant structure where the wave bounces back and forth between reflectors; the frequency of resonance changes in response to environmental changes	Highly sensitive applications like temperature sensing, pressure sensing, and liquid viscosity measurement
One-port SAW resonators 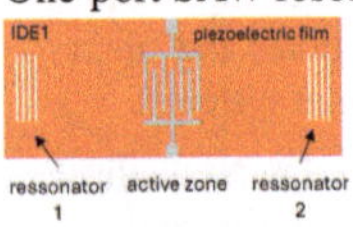	A single transducer acts as both input and output; the wave reflects within a confined area and its characteristics change based on external influences	Used in wireless sensing due to simplicity and efficiency, such as in passive sensing applications
Two-port SAW resonators 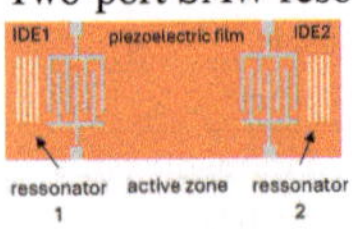	With separate input and output transducers, this configuration offers better control over the sensor's response	Suitable for applications requiring higher sensitivity and more precise measurements, such as RF filters and chemical sensing
Love wave sensors 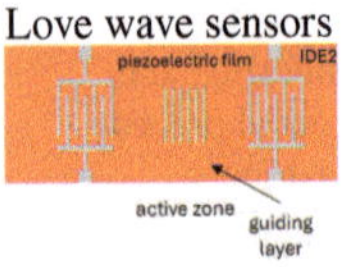	Use Love waves (a wave that travels in a thin guiding layer over the substrate); the wave's characteristics are highly sensitive to surface changes	Ideal for biosensing and chemical sensing in liquids, as they exhibit high sensitivity to mass loading
Shear horizontal SAW (SH-SAW) sensors	Generate shear horizontal waves, which move parallel to the surface. This makes them less sensitive to bulk fluid loading	Used in liquid sensing applications, including biological and chemical sensing, as they operate well in liquid environments
Rayleigh wave SAW sensors	Use Rayleigh waves, which have both vertical and horizontal displacements; these waves are affected by surface changes such as mass, stress, and temperature	Used for gas sensing, temperature sensing, and pressure sensing

these different external stimuli affect the velocity, amplitude, phase, or frequency of the acoustic waves. Primary sensing mechanisms used in SAW sensors are listed in Table 4.17.

Table 4.17 Main sensing mechanisms in SAW sensors

Type	Mechanism	SAW response
Mass loading mechanism	Detects added mass by slowing down the wave. The surface acoustic waves are sensitive to mass changes, Δm, on the sensor's surface. When a mass (e.g., gas molecule, liquid droplet, biological particle) is deposited on the surface of the sensor, it causes a change in the velocity of the acoustic wave due to the added mass. This is the principle of measuring of a QMC—quartz crystal microbalance	Δm results in a decrease in the wave velocity, which shifts its resonance frequency (Δf) or phase. According to Sauerbrey equation (for a quartz crystal): $\Delta f = \frac{2f_0^2}{A\sqrt{\rho\mu}}\Delta m$ f_0—resonant frequency of 1st mode A—Active area ρ—density of piezoelectric material μ—shear modulus of piezoelectric material
Temperature sensitivity mechanism	Measures temperature-induced changes in wave velocity. SAW sensors are sensitive to temperature changes, ΔT, because the velocity of acoustic waves is dependent on the material's physical properties, such as thermal expansion, density, and elasticity	ΔT leads to variations in wave velocity, which can be measured as a shift in resonance frequency or phase. The piezoelectric substrate expands or contracts with temperature, influencing the propagation characteristics
Strain or stress mechanism	Detects mechanical deformation by altering wave propagation. When mechanical strain or stress is applied to a SAW sensor's substrate, it alters the propagation characteristics of the acoustic waves, due to piezoelectric and elastic effects within the substrate material	Stress changes the wave velocity by altering the elastic constants of the material, resulting in a shift in the wave's frequency or phase. The strain can either compress or stretch the piezoelectric material, affecting wave propagation
Acoustic impedance mechanism (liquid sensing)	Sensitive to changes in liquid density and viscosity. In liquid or viscosity sensing, SAW sensors operate based on changes in the acoustic impedance at the interface between the solid substrate and the liquid. When a liquid interacts with the surface of the SAW sensor, the acoustic wave's amplitude, velocity, and attenuation are influenced by the liquid's viscosity and density	The interaction between the acoustic wave and the liquid causes attenuation (wave damping) and a reduction in wave velocity. This can be measured as changes in frequency or amplitude. More viscous or dense liquids increase wave attenuation and shift the resonance frequency

(continued)

Table 4.17 (continued)

Type	Mechanism	SAW response
Dielectric loading mechanism (permittivity sensing)	Detects changes in permittivity or dielectric constant. SAW sensors are also sensitive to changes in the dielectric properties (permittivity) of materials in contact with their surface. Changes in the electric permittivity alter the electromechanical coupling in the piezoelectric substrate, which affects wave propagation	When a material, with a different dielectric constant, contacts the sensor surface, it causes a shift in the wave velocity or phase. The wave's electric field interacts with the dielectric material, influencing its propagation
Viscoelastic sensing mechanism	Sensitive to both mass and mechanical properties of soft materials. When a viscoelastic material (e.g., a polymer or biological sample) interacts with the surface of a SAW sensor, both the mass and the mechanical properties of the material influence the wave propagation	Changes in both the mass loading and mechanical damping of the wave occur. The viscoelastic properties affect the wave velocity and attenuation, which can be detected as shifts in frequency or phase
Piezoelectric perturbation mechanism	Senses electric or mechanical disturbances affecting the piezoelectric substrate. SAW sensors use the piezoelectric properties of their substrate to generate and detect surface acoustic waves. When an external electric field or mechanical stress is applied, the piezoelectric substrate undergoes deformation, altering the wave propagation characteristics	External electric fields or stress change the electric and mechanical properties of the substrate, influencing the wave velocity and phase. This mechanism allows SAW sensors to detect both mechanical and electrical perturbations
Phase shift and time delay mechanism	Measures changes in phase or time delay caused by environmental factors. In some SAW sensors, the sensing mechanism is based on changes in the phase shift or time delay of the acoustic waves as they propagate between the transmitting and receiving transducers. These changes can be due to environmental factors such as temperature, pressure, or mass changes	Any perturbation that affects wave velocity (e.g., mass loading, temperature, or pressure) causes a change in the phase or time delay between the transmitted and received signals. The phase shift or time delay depends on the wave velocity, which is affected by environmental conditions
Frequency shift mechanism (resonant SAW sensors)	Detects external stimuli by observing shifts in the resonant frequency. In resonant SAW sensors, the sensor operates at a specific resonant frequency. When external stimuli (like mass, temperature, or pressure) change the propagation conditions, the resonance frequency shifts	External changes affect the wave velocity, which leads to a measurable frequency shift. This shift is proportional to the intensity of the external stimulus

4.10.3 *Design of SAW Sensors*

Designing a SAW sensor involves optimizing various parameters such as the piezoelectric substrate material, IDE configuration, wavelength, operating frequency, propagation path length, acoustic wave modes, surface modification, and encapsulation. Figure 4.18 shows main geometrical parameters of SAW sensors: finger width, finger spacing/pitch, aperture, number of fingers pairs, substrate thickness, acoustic path length/delay line length, piezoelectric substrate size and shape, reflector design.

- **Piezoelectric substrate material**

The piezoelectric substrate material supports the propagation of surface acoustic waves. It determines the wave velocity, temperature stability, and overall sensitivity of the sensor. Common materials are quartz, lithium niobate ($LiNbO_3$), lithium tantalate ($LiTaO_3$), zinc oxide (ZnO). These piezoelectric materials show high piezoelectric constants but are very brittle, limiting their use in flexible devices. Polymer-based piezoelectric materials (Sect. 4.10.5) are more flexible but have lower piezoelectric constants. The SAW sensor also needs to maintain consistent performance over a range of temperatures. Substrate material impacts temperature stability. Materials like quartz offer good temperature stability, while others like lithium niobate may require temperature compensation techniques. IDE are to be printed over the piezoelectric substrate and good compatibility between substrate and ink shall be also guaranteed.

- **IDE configuration**

IDE are metal electrodes patterned on the substrate to generate (transmitter, Tx) and detect (receiver, Rx) acoustic waves. Their main geometric parameters are described in Table 4.18 (see Fig. 4.18): the finger width, W (printed line width), the finger spacing, G, the number of fingers, N, the aperture, A (width of the transducer perpendicular to the direction of wave propagation). The pitch refers to the distance between adjacent fingers in the IDE and determines the wavelength of the surface acoustic

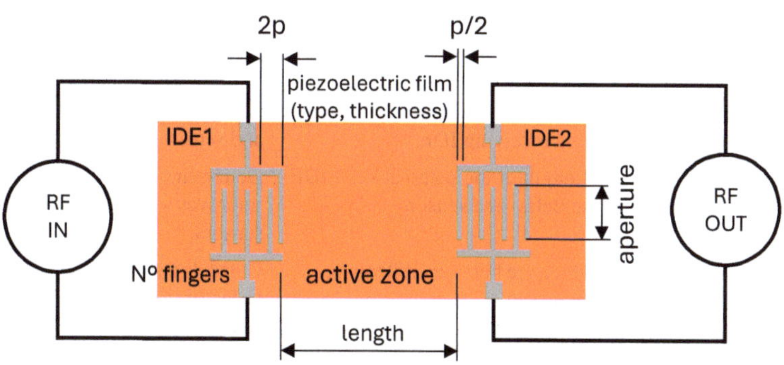

Fig. 4.18 Main design parameters of SAW sensors (p is the IDE pitch)

wave generated and, consequently, the frequency at which the sensor operates. The pitch of the IDE, p, is half of the wavelength, λ, of the acoustic wave that propagates across the surface of the sensor: $\lambda = 2p$. Also, typically $W = G = p/2$. The settings of these parameters affect the frequency, wavelength, and efficiency of wave generation and detection.

- **Wavelength**

Wavelength (λ) is the distance between successive peaks of the acoustic wave, being determined by geometric parameters:

$$\lambda = 2(W + G). \tag{4.26}$$

The wavelength determines the operating frequency and resolution. Shorter wavelengths lead to higher frequencies and better resolution.

- **Operating frequency**

The operating frequency, f, is the frequency at which the SAW device operates:

$$f = \lambda \cdot v \tag{4.27}$$

Table 4.18 Main geometrical parameters of SAW sensors (see Fig. 4.18)

Geometrical parameter	Description	Impact on SAW sensor performance
Finger width (W)	Width of individual IDE fingers, typically $\lambda/4$	Affects the efficiency of wave generation and resonant frequency
Finger spacing (S)	Gap between adjacent IDE fingers, typically $\lambda/4$	Controls wavelength and operating frequency
N° of finger pairs (N)	Number of interdigitated finger pairs in the IDE	Higher N increases wave energy and sensitivity
Aperture (A)	Width of the IDE perpendicular to wave propagation	Larger aperture increases energy transfer and sensitivity
Substrate thickness (T)	Thickness of the piezoelectric substrate	Affects mechanical stability and wave propagation
Acoustic path length (L)	Distance between transmitting and receiving IDE	Longer path increases time delay and sensitivity
Delay line length	Length of the path between IDE in delay line sensors	Determines time delay and improves sensitivity in delay line sensors
Reflector spacing (D)	Distance between reflectors, typically $\lambda/2$	Controls wave reflection, Q-factor, and frequency selectivity

where v is the velocity of the surface acoustic wave in the substrate material. The frequency (and wavelength) of the generated surface acoustic waves is determined by the spacing of the fingers in the IDE and the velocity of the wave in the piezoelectric material. Higher frequencies provide better sensitivity and resolution but may require more precise fabrication techniques. A balance between wavelength and IDE configuration is required to achieve the desired operating frequency.

- **Propagation path length**

The propagation path length, L, is the distance the acoustic wave travels between the transmitting and receiving IDE. Longer paths can increase sensitivity but may also increase attenuation and delay.

- **Acoustic wave mode**

The acoustic wave mode is the type of acoustic wave used in the sensor. Common modes are: Rayleigh waves, Love waves, and Shear Horizontal waves, presented in Fig. 4.19.

Different modes are suited to different applications. Rayleigh waves are common in many SAW devices. They involve particle motion both parallel and perpendicular to the surface, being sensitive to surface changes. Rayleigh waves are sensitive to both vertical and horizontal deformations, making them useful for detecting both compressional and shear deformations. They are widely used in gas and pressure sensors. Love waves are used for liquid sensing due to their confinement to the surface. They are also highly sensitive to shear deformations, making them useful in applications where shear horizontal motion is important. Shear horizontal surface acoustic waves (SH-SAW) involve horizontal motion parallel to the surface and perpendicular to the direction of wave propagation (i.e., horizontal motion with no vertical component). They are especially useful for sensing in liquid environments because they are less affected by liquid damping. Table 4.19 compares these types of acoustic wave modes.

- **Active layer**

The surface modification of the sensor active layer with specific coatings or functional groups aims at targeting particular analytes. Functionalization of the sensor surface improves selectivity and sensitivity. Coatings with polymers, metals, or biological

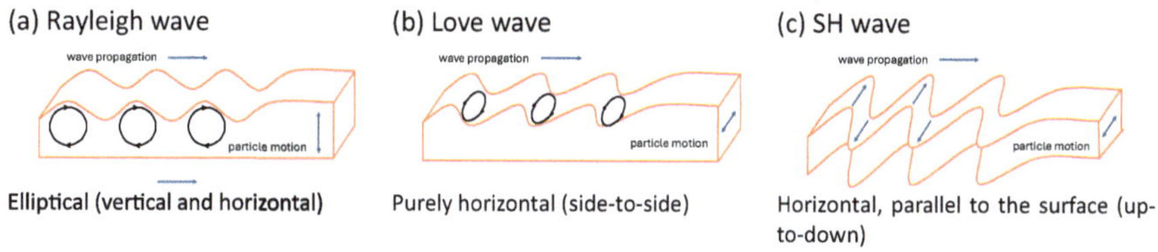

Fig. 4.19 Types of surfaces waves in SAW sensors

Table 4.19 Comparison between acoustic wave modes in SAW sensors

Feature	Rayleigh waves	Love waves	SH waves
Nature	Surface acoustic wave (SAW) that travels along the surface of an elastic solid	Also surface acoustic wave, but they require a layered structure	Shear waves where the particle motion is horizontal and parallel to the surface but perpendicular to the propagation direction
Particle motion	Elliptical (vertical and longitudinal)	Horizontal (perpendicular to propagation)	Horizontal (parallel to surface, transverse)
Propagation medium	Surface of an elastic solid	Requires a soft layer over a hard substrate	Surface or bulk (depending on mode)
Depth sensitivity	Exponentially decays with depth	Confined to the top layer	Variable; depends on mode
Sensitivity	Sensitive to surface conditions	Highly sensitive to surface conditions	Less sensitive to surface, ideal in liquids
Common applications	Environmental and material sensors	Biochemical and surface-bound sensing	Liquid-phase biosensors, chemical sensing

recognition elements are common to enhance the interaction with specific analytes, improving the sensor's performance for targeted applications.

- **Encapsulation**

Encapsulation of SAW sensors is a critical design step that protects the sensor from environmental factors (such as humidity, dust, contaminants, chemicals and mechanical impacts) while preserving its sensitivity to the target stimuli, i.e., without interfering with the propagation of surface acoustic waves. Effective encapsulation enhances sensor durability, performance, and longevity. Encapsulation of SAW sensors is a balance between protection and sensitivity. The choice of encapsulation method and material depends heavily on the sensor's application environment. Encapsulation materials are chosen based on their chemical inertness, mechanical stability, and compatibility with the sensing application. Common materials include glass, ceramics, and specially designed polymers that protect the sensor while allowing its surface to interact with the environment if necessary. For flexible applications, the SAW sensor needs to conform to curved surfaces, to stretch, or to bend, and the encapsulation material needs to be not only protective but also highly flexible and resilient. The best materials suited for encapsulating flexible SAW sensors are PDMS, PU, silicone rubber, TPU, PI, and Parylene coatings.

4.10.4 Key Parameters in Printed Flexible SAW Sensors

SAW sensors have a range of performance metrics used to evaluate their effectiveness in various applications, namely;

Sensitivity—measures how effectively the SAW sensor responds to changes in the parameter being measured. Often measured as a frequency shift per unit change in the sensing parameter (e.g., Hz/°C for temperature or Hz/ppm for gas concentration). High sensitivity allows the sensor to detect small changes, making it useful for precision sensing in various applications.

Resolution—is the smallest change in the sensing parameter that the SAW sensor can detect, being dependent on the sensing application (e.g., ppm for gas sensors or °C for temperature sensors). Higher resolution means the sensor can detect finer details in the measured parameter (high precision).

Dynamic range—is the response range over which the SAW sensor can accurately measure the sensing parameter. It is often expressed in decibels (dB) or as a ratio between the smallest and largest detectable values. A wide dynamic range is necessary for sensors used in environments where the measured parameter varies significantly.

Response time—is the time the SAW sensor takes to respond to a change in the measured parameter. Fast response time is critical for real-time monitoring and applications that require rapid detection, such as gas leak detection or instantaneous pressure changes.

Frequency shift—is the change in the resonant frequency of the SAW sensor when the sensing parameter changes. The magnitude of the frequency shift is often used to quantify the sensor's response to the external environment. A larger frequency shift generally indicates higher sensitivity.

Stability—is the ability of the SAW sensor to maintain consistent performance over time and under varying environmental conditions. A stable sensor will have minimal drift in output over time, ensuring accurate and reliable measurements.

Temperature coefficient of frequency (TCF)—is the rate at which the resonant frequency of the SAW sensor changes with temperature, measured in Hz/°C. Minimizing the TCF is important for applications where temperature variations are expected, as large frequency shifts due to temperature can affect the accuracy of the sensor. SAW sensors often require temperature compensation techniques to maintain accuracy.

Noise level—is the inherent electrical noise in the sensor's output, which can obscure small signals and affect the accuracy of the sensor. It is measured in decibels (dB) or as a signal-to-noise ratio (SNR). Low noise levels are critical for detecting small changes in the sensing parameter, especially in high-sensitivity applications.

Power consumption—is the amount of electrical power required by the SAW sensor to operate, measured in watts or milliwatts. Low-power consumption is important for battery-operated and portable applications, where energy efficiency is a priority.

Long-term drift—is the gradual change in the sensor's baseline output over a long period, even when the measured parameter is constant. Minimizing drift is essential for applications that require long-term stability and reliability.

Selectivity—is the ability of the SAW sensor to distinguish between the target substance or parameter and other interfering substances or environmental factors. High selectivity is critical in chemical and biosensing applications, where the sensor needs to respond specifically to the target substance while ignoring others.

Linearity—is the degree to which the sensor's output is directly proportional to the measured parameter across the sensing range. A linear response simplifies calibration and data interpretation, which is particularly important in precision sensing applications.

Operating temperature range—is the range of temperatures over which the SAW sensor can function properly without significant degradation in performance. A wide operating temperature range is important for sensors used in harsh or variable environmental conditions.

4.10.5 Materials for Printed Flexible SAW Sensors

Flexible materials used in SAW devices are important for applications in wearable technology, flexible electronics, and sensors that require conformability to non-flat surfaces. Main flexible materials that can be used in printed flexible SAW sensors are:

- A **flexible substrate** is the base layer on which the SAW device is fabricated. They allow the SAW device to be bent, stretched, or conformed to different surfaces. Typically, they are polymeric substrates like PI, PDMS, PET, PEN, TPU.
- Flexible **piezoelectric materials** are essential for generating and propagating acoustic waves in SAW devices. Flexible piezoelectric materials allow SAW devices to maintain piezoelectric properties while being bent or stretched. Table 4.20 compares the main properties of flexible and rigid piezoelectrical materials typically used in SAW sensors. The former includes polymeric materials such as PVDF and their copolymers (e.g., PVdF-TrFE), polylactic acid (PLA), collagen, piezoelectric ceramics (PZT, $BaTiO_3$), and polymer-based composites.
- Flexible **conductive materials** are required for printing IDE. These include flexible conductive inks, carbon allotropes (graphene, CNT), silver nanowires, conductive polymers (PEDOT:PSS).
- Flexible **encapsulation materials** for protecting flexible SAW devices from environmental damage (humidity, chemicals, dust) while maintaining flexibility includes PDMS and other silicones, PU, parylenes.
- Flexible **dielectric materials** are important in SAW devices to control the propagation of the surface acoustic waves and can serve as protective layers, including materials such as fluoropolymers (e.g., Cytop®) and acrylic-based coatings.

Table 4.20 Comparison of properties of rigid and flexible piezoelectric materials

Piezoelectric material	d_{33}/d_{31} (pC/N)	ε_r	E (GPa)	ε_b (%)
Rigid				
Quartz (SiO_2)	3–5/1–3	4–5	70–97	0.1–0.2
PZT	200–600/80–250	1000–3000	60–80	0.1–0.2
Lithium niobate ($LiNbO_3$)	6–25/8	30	70–110	0.2
Lithium tantalate ($LiTaO_3$)	5–18/1–8	54	50	0.2
Barium titanate ($BaTiO_3$)	100–200/150	1200	67	0.2–0.5
Zinc oxide (ZnO)	5–12/11	12	210	0.1–0.2
Aluminum nitrite (AIN)	2–6/2	8–10	330–280	0.1–0.2
Flexible				
PVdF	10–20/20–30	5–20	2	25–500
PVdF-TrFEE	10–64/-	30–60	1–2	10–300
PZT composite*	50–400/20–150	100–1000	5–20	10–20
$BaTiO_3$ composite*	10–40/-	15–60	1–5	10–20
AlScN thin film**	5–10/2–4	9–10	350–400	0.1–0.2

d_{33}/d_{31}—Piezoelectric coefficients, ε_r—dielectric constant, E—Young´s Modulus, ε_b—Strain at break; Higher values of d_{33} and d_{31} results in better energy conversion efficiency. Higher dielectric constant leads to better energy conversion efficiency; *composite properties depend upon the percentage of incorporation of nanoparticles in the composite and upon the matrix. **thin films of rigid materials like AlScN or ZnO can handle slightly higher strain due to confinement effects or improved substrate support

Chapter 5
Recent Advances, Challenges, and Future Trends

Printed flexible electronics technology is steadily reshaping the future of electronics by:

- redesigning form factors with thin, bendable, stretchable, and lightweight designs;
- enabling ubiquitous electronics, leveraging smart connected products: "smart everything and everywhere";
- disrupting manufacturing paradigms by adopting additive manufacturing techniques, reducing waste and energy consumption, and lowering costs; and
- supporting sustainability goals by easier recycling, by using greener materials and low-energy manufacturing techniques, toward greener technologies.

PFSs have emerged at the forefront of this evolution, fueled by the development of novel functional materials, high-resolution and additive printing techniques, and increasingly seamless system-level integration strategies. These advances are not only enhancing device performance but also expanding the horizons of PFS applications across diverse sectors, empowering cross-disciplinary innovations and cross-sector integration, opening the door to entirely new use sensing solutions [11].

The field of printed flexible electronics is evolving rapidly, marked by innovations that address the challenges of flexibility, device reliability, manufacturing scalability, durability, and sustainability. The technological challenges of PFS in terms of materials and manufacturing are relevant and addressing them will be imperative for their widespread adoption, unlocking their full potential in applications.

Despite these promising developments, critical technological hurdles persist—particularly in the realms of material stability, electrical conductivity under strain, multilayer device architectures, and scalable, reproducible and cost-effective fabrication. Addressing these challenges will be pivotal in enabling the transition from proof-of-concept prototypes to robust, commercially viable solutions that can be deployed at large scale.

Looking ahead, the true potential of PFS lies in their convergence with other emerging technologies such as the Internet of Things (IoT), artificial intelligence

J. C. Viana, *Primer on Printed Flexible Sensors*,
https://doi.org/10.1007/978-3-032-06554-4_5

(AI), soft robotics, next-generation wearable devices, and personalized healthcare. This interdisciplinary integration will catalyze groundbreaking innovations—such as self-powered health monitors, intelligent packaging, adaptive human–machine interfaces, and ultra-lightweight sensor networks—paving the way for a new era of ubiquitous, smart, and sustainable electronics.

5.1 Technological Challenges (Materials and Manufacturing)

The development and commercialization of PFS face several technological challenges, particularly in the areas of materials and manufacturing. Addressing these challenges is fundamental for achieving reliable and durable, high-performance sensors that can be produced cost-effectively at large volumes.

5.1.1 Materials Innovations

Material innovations in PFS are related to the enhancement of their performance and electrical conductivity under strain, their stability and durability under the application conditions, and sustainability issues. Main challenges are focused on the development of new materials (substrates, inks) with better properties, such as stretchable conductive materials with low and constant conductivity, hybrid nanomaterials, e.g., combining graphene with metallic nanoparticles for improved conductivity while keeping transparency, self-healing and self-reparable materials, biocompatible materials, biodegradable and low-environmental impact materials. Main challenges in material innovations for PFS are:

- **Oxidation-resistant conductive inks**

Traditional conductive inks like the ones based on Ag and Cu nanoparticles, while offering high electrical conductivity, can suffer from oxidation, leading to reduced performance over time. This oxidation is particularly problematic in humid or harsh environmental conditions, impacting sensor longevity and reliability. Recent advances in nanomaterials like graphene, CNT, MXenes, and metallic nanowires have improved the sensitivity, flexibility, and electrical conductivity of printed sensors. These materials allow the production of highly responsive sensors that can detect very small changes in stimuli (e.g., pressure, temperature, humidity, biochemical markers).

- **Stretchable substrates and inks**

The development of stretchable inks containing Ag nanoparticles, conductive polymers (e.g., PEDOTPSS), or liquid metal alloys has enabled the production of sensors that maintain functionality even under extreme stretching and bending. While materials like CNT and graphene offer excellent flexibility, integrating them into conductive inks without compromising their properties is complex. Flexible substrates need to balance flexibility, robustness, and durability. Many existing substrates, such as PET and PEN, can still be prone to mechanical fatigue, limiting sensor lifespan. Ensuring that conductive materials remain functional under repeated bending, stretching, or twisting is also challenging.

- **Low-temperature curing inks/High-temperature-resistant substrates**

Some functional inks require elevated temperatures to cure, which can deform or degrade low-cost flexible substrates. This restricts the range of substrates that can be used, particularly in applications requiring high-temperature processing. It is therefore required to explore low-temperature curing inks, including those that cure via photonic sintering, UV, or chemical processes. Resolving this challenge is critical for enabling multilayer integration, improving device reliability, and ensuring manufacturability on large-scale roll-to-roll printing platforms.

- **Biocompatible and biodegradable inks**

For wearable and biomedical applications, the materials used need to be biocompatible, non-toxic, and safe for prolonged skin contact, which limits the choice of conductive inks and substrates. For wearable or medical applications, substrates must be biocompatible and non-toxic. Additionally, there is growing interest in developing biodegradable or eco-friendly substrates to reduce electronic waste, but these materials often have lower performance compared to synthetic polymers. Their long-term behavior is also relevant, enlarging their applications rather than disposable or short-term usage.

- **Strong adhesion between inks-substrates**

Ensuring strong adhesion between different materials (e.g., conductive inks and flexible substrates) is critical for sensor functionality, reliability, especially during repeated bending or stretching. Poor adhesion can lead to cracking or delamination under mechanical stress and therefore to limited durability.

- **New ink formulations for consistent manufacturing**

Ensuring that inks have the right viscosity, flow properties, and drying characteristics for different printing techniques is fundamental for consistent manufacturing. However, achieving optimal formulation can be challenging, especially when dealing with new materials like nanomaterials or conductive polymers. Furthermore, achieving optimal interaction between functional inks and flexible substrates is critical. Variations in surface energy, roughness, and porosity affect ink adhesion, spreading, and curing behavior. Optimizing simultaneously ink-substrate-printing techniques is vital for superior device performance.

- **New encapsulation materials**

Flexible sensors are often exposed to humidity, UV radiation, chemicals, and mechanical abrasion, which can degrade their performance. Finding suitable encapsulation materials that protect the sensor while maintaining flexibility is challenging. Encapsulation layers can sometimes reduce the sensitivity of the sensor to the target stimuli. Balancing protection with sensor performance is critical. Developing materials with stable long-term performance is a key challenge.

- **Self-healing materials**

Self-healing sensors can repair themselves when damaged, restoring functionality without needing manual intervention. It is still a challenge to develop self-healing polymers, conductive hydrogels, and liquid metal inks that can restore electrical pathways after damage.

5.1.2 Manufacturing Innovations

Printing technologies are central to flexible electronics on a large scale. Conventional (screen, inkjet) and emergent (aerosol jet, electrohydrodynamic) printing methods must leverage the compatibility of flexible substrates with low-temperature curing inks. However, to further enhance the performance and scalability of these printing techniques, several challenges must be addressed. These include limitations in printing resolution, in the efficient use of materials, in ink-substrate compatibility, in the precise alignment in multilayer structures, in the uniformity in film thickness, in scalability of high-throughput processes, and in ensuring long-term reliability under mechanical and environmental stress. Successfully addressing these challenges is essential for achieving the full potential of PFS in a wide range of industrial and consumer applications. Main challenges in manufacturing innovations for PFS are:

- **Improvements in printing technologies**

Improvements in printing technologies, like inkjet printing, R2R printing, aerosol jet printing, and electrohydrodynamic printing, can provide better resolution, scalability, and material compatibility. Achieving high-resolution printing with uniform thickness and consistency is essential for reliable sensor performance. However, maintaining this precision over large areas or complex surfaces can be difficult, especially when using techniques like screen printing or inkjet printing. In fact, inkjet, screen, and gravure printing techniques present some limitations. Each printing technology has its limits in terms of resolution, speed, and layer thickness control. Inconsistent ink flow, smearing, and substrate wrinkling during the printing process can impact device quality. Also, precise alignment between layers during multilayer printing is also a challenge, impacting device yield and consistency.

Many conductive inks require curing or sintering at elevated temperatures, which can damage or distort flexible substrates. Developing low-temperature processing techniques is essential to maintain substrate integrity. Techniques like photonic curing and laser sintering allow low-temperature processing of conductive inks, making it possible to use a wider range of flexible substrates without damage.

Transfer printing techniques have also made strides, enabling the transfer of nanoscale materials and devices onto flexible substrates. This approach is essential for fabricating high-performance flexible electronics but requires careful control of stamping processes to minimize defects. Multilayer printing is often required for complex sensors, necessitating precise alignment between layers. Misalignment can lead to performance issues or even sensor failure.

- **Scaling-up printing technologies**

Scaling up from lab-scale printing to industrial-scale manufacturing while maintaining quality and performance is a significant challenge. This is particularly true for advanced materials like graphene and CNT, which may require precise deposition techniques. In scaling-up printing processes keeping quality and consistency of PFS is necessary. Besides, high-speed printing methods are needed to make manufacturing cost-effective, but increasing speed can compromise precision and quality.

- **New printing techniques**

High-resolution 3D printing and advanced inkjet printing techniques have enabled the fabrication of complex sensor geometries with precise control over material deposition, enhancing sensor performance. Direct printing on complex surfaces is also a challenge. Printing sensors directly onto textiles, curved surfaces, or human skin is now possible, expanding the potential applications of flexible sensors in wearables, robotics, and smart packaging.

Flexible sensors often require protective encapsulation or packaging, which must be done without compromising flexibility, sensitivity, or performance. Traditional packaging techniques are not always suitable for flexible devices, necessitating the development of new methods.

5.1.3 Integration Issues

The integration of PFS into devices presents several challenges from materials and manufacturing, mechanical design, connectors and interface electronics, and environmental issues [12]. Materials compatibility is a challenge (e.g., ink-substrate adhesion, thermal and mechanical properties). Scaling and consistency of manufacturing processes are important challenges for large-scale and cost-effective production. Also, the combination of traditional rigid electronics with PFS may pose integration issues. Overcoming these integration barriers is essential to ensure the reliable performance of the next generation of sensing systems.

The thermal mismatch between materials (substrate, ink, encapsulant, rigid components) leads to different expansions or contractions when exposed to temperature changes that may lead to delamination or cracks. Ensuring strong adhesion between flexible and rigid parts of a device can be difficult, particularly when different material types are involved. Flexible devices often need to be connected to rigid systems (e.g., batteries, processors), and maintaining reliable electrical connections between flexible and rigid parts is challenging due to stress at the junctions. Printed flexible devices typically operate at lower temperatures, and managing heat dissipation in flexible electronics is more complex than in rigid systems. Flexible substrates (e.g., some polymers like PET with low glass transition temperatures) and inks can deteriorate at lower temperatures than traditional rigid components, which limits the operational temperature range of the devices. High-temperature-resistant substrates can be used (like PI, PDMS). Flexible materials have lower thermal conductivity compared to metals or ceramics, which can be problematic for heat dissipation in power-intensive applications.

The process of printing flexible devices and integrating them into complex systems involves scaling production while maintaining quality and consistency. Achieving high-resolution, consistent printing across large areas can be challenging, particularly for fine conductive lines or thin films, leading to defects or inconsistencies in performance. Many flexible devices require multiple layers (e.g., conductive, dielectric, piezoelectric), and aligning these layers accurately during printing is difficult. While printing techniques like R2R and inkjet printing are cost-effective, ensuring uniformity and defect-free production in large-area and high-scale production remains a challenge.

Many flexible devices need to interface with traditional rigid electronics (e.g., processors, communication modules) for enhanced functionality, reliability, and signal processing capabilities, requiring reliable interconnections between flexible and rigid components (hybrid integration). Establishing reliable electrical connections between flexible and rigid elements, handling the mechanical mismatch between them, while maintaining overall flexibility is a complex manufacturing challenge. Traditional soldering techniques may not work well with flexible substrates. The field of printed flexible electronics is still evolving, and there is a lack of standardized design and manufacturing processes, making integration into larger systems complex. Printed flexible electronics lack standardized design rules and manufacturing protocols, making it harder to ensure interoperability and integration across different devices. While printed flexible electronics can potentially reduce costs in the long term, the initial setup, R&D, and low-yield production can make integration expensive. The cost of developing and prototyping new materials, processes, and integration techniques can be high, especially for custom flexible devices. Achieving high manufacturing yield with minimal defects is difficult, and early-stage manufacturing may result in higher costs.

5.1.4 New Sensors

The next generation of PFS will be driven by breakthroughs in materials science, advanced printing techniques, and innovative device design, significantly expanding their functionalities and applications. Emerging multifunctional sensors can detect multiple stimuli within a single, integrated flexible platform. Additionally, as already referred to, stretchable and ultra-flexible sensors are being engineered to maintain stable performance under extreme mechanical deformation.

Another exciting direction is the development of biodegradable and transient sensors, which are designed to degrade safely and naturally over time. The use of bioderived and biodegradable polymers, cellulose-based substrates, and organic inks, and the reuse of recycled materials, combined with eco-friendly manufacturing processes is fundamental for superior sustainability.

Flexible energy-harvesting sensors detect not only environmental or physiological changes but also generate their own electrical power by harvesting energy from mechanical, thermal, or solar sources. The integration of energy-harvesting techniques such as piezoelectric, triboelectric, and thermoelectric energy harvesting, enables self-powered sensor operation, reducing the need for external batteries or prolonging their lifetime. These devices will be key for wearable sensors without the need for external batteries. The generation of power autonomously will also impact the widespread of IoT devices.

Incorporating wireless communication modules like Bluetooth, NFC, and RFID into flexible sensors will facilitate the development of real-time and remote monitoring systems.

Printable 3D flexible sensors have more complex forms and functionalities, allowing them to be integrated into non-planar or curved surfaces (e.g., 3D-printed wearable sensors, 3D-conformal sensors). Flexible optical sensors detect light, color, or optical signals using flexible, printed materials (e.g., wearable UV detectors, optical imaging sensors). They will be based on organic photodetectors, quantum dot inks, and flexible OLED materials to create fully flexible optical sensors.

Electronic devices and sensors can be fabricated as thin, flexible patches or "electronic tattoos" that adhere to the skin's surface like temporary tattoos. These tattoo electronics can monitor physiological signals, such as muscle activity, brain waves, or biochemical markers, for applications in healthcare, human–computer interaction, and sports performance monitoring. Skin-mounted sensors can provide tactile feedback or haptic sensations to enhance augmented reality (AR) experiences. By integrating actuators or vibrotactile transducers directly onto the skin, AR interfaces can deliver immersive sensory feedback to the wearer, such as virtual touch, texture, or spatial cues, enhancing the realism and interactivity of virtual environments. Skin-mounted sensors can be integrated into prosthetic limbs, orthotic devices, exoskeletons, or assistive technologies to provide sensory feedback, motion detection, and control functionalities. These devices can enhance the mobility, dexterity, and autonomy of individuals with disabilities or limb loss, enabling them to interact with their environment more effectively. Collectively, these innovations are positioning the next wave of PFS as highly deformable, adaptable, multifunctional, and eco-conscious technologies that will play a central role in shaping the future of smart connected systems.

5.1.5 Integration of Artificial Intelligence and Machine Learning

Using artificial intelligence (AI) for real-time quality control and predictive maintenance can help detect manufacturing defects early, ensuring consistent quality and reducing waste. This opens new possibilities for smart, adaptive, and self-learning systems. Combining sensor technology with AI/ML (machine learning) allows for more intelligent data processing, better decision-making, the ability to detect complex patterns allowing predictive analytics, personalized sensing, and autonomous system control.

ML algorithms can be trained to detect when a sensor has drifted or is behaving abnormally, allowing for self-calibration. By continuously learning from the data, the system can adapt its model and recalibrate the sensors in real-time without the need for manual intervention. ML models can be trained on historical sensor data to predict when a sensor is likely to fail or degrade. This enables predictive maintenance, where systems are monitored continuously to detect early signs of failure and prompt repairs or replacements before a malfunction occurs. ML algorithms can create personalized models for users based on individual-specific data, offering customized feedback and recommendations. The system learns over time, improving its accuracy and relevance for each user. Sensor fusion algorithms combine data from multiple types of sensors (e.g., temperature, humidity, and pressure sensors) to provide a more comprehensive understanding of the environment. ML can identify correlations between different data streams to enhance overall system accuracy and functionality. ML algorithms, such as unsupervised learning models, can be used for anomaly detection, automatically identifying when sensor data deviates from expected patterns. This enables systems to detect faulty sensors or dangerous conditions early. AI and ML models enable devices to learn and adapt over time to the specific conditions and preferences of individual users or environments, allowing for context-aware sensing. AI models integrated with flexible sensors allow robots and autonomous systems to process real-time data, make decisions, and adapt their behavior autonomously, allowing them to interact with its environment more naturally and effectively.

5.2 Performance, Reliability, and Durability

Achieving high performance and reliability is paramount for the widespread adoption of PFS. They must deliver dependable performance under varying mechanical stresses, environmental conditions, and usage cycles. Key challenges remain, such as enhancing sensitivity and selectivity, improving mechanical durability under repeated bending or stretching, and ensuring stability against factors like humidity, temperature fluctuations, and chemical exposure. As these technologies mature, PFSs are becoming increasingly viable for demanding, real-world applications. In the medical field, for example, reliable and accurate flexible sensors can enable continuous monitoring of vital signs with clinical-grade precision. In industrial environments, robust flexible systems can be deployed on curved or dynamic surfaces for structural health monitoring or predictive maintenance. Ultimately, by addressing these performance, reliability, and durability challenges, PFS will be recognized as trusted components in critical systems, accelerating their integration into everyday life and supporting the development of smarter, more responsive technologies across sectors.

5.2.1 Performance

The critical assessment of the performance of PFS can be made based on their more relevant FOM, shown in Table 5.1. Challenges such as reproducibility, scalability, integration with electronics, and long-term reliability are of paramount importance.

5.2.2 Reliability and Durability

Reliability factors of PFS are related to their mechanical durability, environmental stability, aging and fatigue, and electrical reliability. PFS must maintain functionality under repeated mechanical stress, e.g., bending, stretching, and twisting. The ability to withstand thousands or even millions of bending cycles is essential for applications in wearables or flexible electronics. Mechanical fatigue is an important property for PFS that can lead to degradation of sensor performance over time. Ensuring that flexible sensors maintain functionality under mechanical stress remains a challenge. Also, conductive inks and substrates can crack, delaminate, or lose conductivity over time, leading to reduced sensor performance. Ensuring long-term durability under mechanical stress is one of the most significant reliability challenges.

Flexible sensors may be exposed to varying temperature and humidity levels, which can affect their performance. Changes in environmental conditions can lead to shifts in sensor readings (drift) or even permanent damage. In some applications, sensors may encounter chemicals or UV radiation that could degrade the sensor materials or reduce their accuracy.

Over time, materials used in flexible sensors can degrade (e.g., aging and fatigue), especially when exposed to harsh conditions. This degradation can lead to changes in sensor properties, affecting performance. Long-term stability of FPS, maintaining consistent performance over time, despite repeated use or exposure to various conditions, is a significant reliability challenge.

Table 5.1 Main challenges on the performance of PFS

Type	Description	Challenges
Sensitivity	Ability to detect small changes in the measured parameter. It is affected by the choice of functional inks, substrate properties, sensor design, and the printing process	Achieving high sensitivity while maintaining flexibility is challenging, as traditional methods to increase sensitivity (e.g., thicker sensing layers) can compromise flexibility
Resolution and range	Resolution refers to the smallest change the sensor can detect; the range defines the span of values the sensor can measure accurately	Materials limit extreme performance, inconsistencies and sensitivity to external factors (e.g., temperature, humidity), impacting accuracy; limited fabrication precision affects resolution
Response time	Is the time that the sensor takes to detect a change and produce a measurable output. It is influenced by the material properties, its thickness, and the interactions between the sensing element and the target stimulus	Flexible sensors sometimes have slower response times (e.g., slower electron mobility, diffusion-based mechanisms) than rigid ones due to delays in signal transmission through softer materials, printed layer thickness variations and imperfect interfaces
Linearity	Refers to the extent to which the sensor output is directly proportional to the measured parameter	Nonlinearities can arise due to uneven material distribution, substrate deformation, or inconsistent printing, making calibration more complex
Signal-to-Noise Ratio (SNR)	Measures the strength of the sensor's signal relative to background noise. A high snr is essential for accurate readings, particularly in low-signal applications	Flexible sensors often have lower snr compared to rigid sensors due to the variability in printed conductive paths, which can introduce noise or inconsistencies
Stability	Refers to its ability to maintain consistent and reliable performance over time under varying operational and environmental conditions, without significant drift or degradation	Flexible sensors often have lower signal strength, leading to challenges in signal processing and accuracy; sensor stability is affected by environmental factors and mechanical fatigue; materials degradation (aging, wear) shorten sensor lifespan

Ensuring that the electrical signals generated by the PFS remain consistent, even after repeated mechanical deformation or harsh conditions, is critical for their reliable performance. Maintaining stable electrical connections between flexible components, especially when integrating with rigid electronics, is essential for preventing signal loss or degradation.

5.2.3 *Strategies for Improving Performance and Reliability*

Using stretchable conductive inks (based on Ag nanowires, CNT, graphene, liquid metals) can improve flexibility and durability while maintaining electrical conductivity. Combining different materials (e.g., conductive polymers with nanoparticles) can enhance sensitivity, stretchability, and environmental stability of PFS. Furthermore, recent advances in self-healing materials (e.g., vitrimers, poly(urethane-urea) elastomers, ionomers, shape memory polymers) enable flexible sensors to recover from minor damages or cracks, improving longevity and reliability.

High-resolution printing can improve the performance and reliability of PFS. Using advanced printing methods like aerosol jet or electrohydrodynamic printing allows for precise deposition of materials, improving sensor performance and reducing defects. Creating multilayer sensors can enhance performance by integrating different sensing elements, protective layers, and conductive paths, improving sensitivity and reliability.

Applying protective coatings or encapsulating the sensor in a flexible material provides environmental protection, shields it from humidity, chemicals, or UV radiation, enhancing durability without compromising flexibility. Using flexible reinforcement layers can improve mechanical durability, reducing the likelihood of cracks or delamination under repeated stress.

A smart design of the sensor through its structural optimization is fundamental. Designing sensor patterns that distribute mechanical stress more evenly can improve durability. For example, serpentine or mesh-like structures can absorb strain more effectively.

Variability in material properties, printing techniques, and environmental conditions can lead to inconsistencies in sensor performance, making it difficult to ensure high-quality, uniform production. The development and implementation of quality control techniques is required.

5.3 Scalability and Cost-Effectiveness

The scalability and cost-effectiveness of PFS are improving due to material innovations, advances in printing techniques, and process optimizations. By adopting high-speed manufacturing methods like R2R printing, using low-cost, recyclable conductive materials, and implementing automation, PFS can be produced at scale while keeping costs manageable. These developments are paving the way for PFS to be integrated into a wide range of applications, making them an integral part of the future of sensor technology.

5.3.1 *Scalability Challenges*

- **Printing techniques and equipment**

Many advanced printing techniques used for flexible sensors, such as inkjet and screen printing, can be relatively slow when producing high-resolution, multilayered designs. This limits production speed and throughput, making it challenging to scale up for large-volume manufacturing. Furthermore, maintaining precision and uniformity in sensor performance across large production volumes is difficult, particularly when dealing with nanomaterials-based inks or complex sensor designs. Variations in layer thickness, alignment, or material distribution can impact the quality and consistency of sensors.

- **Material deposition and wastage**

Many printing processes have relatively low material use efficiency, leading to wastage of expensive conductive inks or nanomaterials. This inefficiency not only increases costs but also makes it harder to achieve scalable production. Material efficiency is an important challenge. Also, printing multiple layers of different materials (e.g., conductive, insulating, and sensing layers) can lead to adhesion issues, alignment challenges, and increased processing times, especially when scaling-up production.

- **Substrate handling and rigid components integration**

Handling large, flexible substrates during the manufacturing process can be challenging, as they are prone to deformation, wrinkling, or stretching. This can lead to defects, misalignment, and reduced yield when producing sensors at scale.

Integrating flexible sensors with other electronic components, such as rigid microchips, antennas, or batteries, on a large scale, requires precise alignment and bonding, which can be time-consuming and complex.

- **Scalability and manufacturing challenges**

Maintaining high precision and uniformity in large-scale production is challenging. Implementing quality control and testing procedures is needed for detecting defects in PFS. Quality control is more challenging for PFS than rigid ones because defects like cracks, delamination, or uneven printing may only appear under certain mechanical stresses. Traditional testing methods might not be adequate for detecting these issues. Creating industry standards for materials, processes, and performance metrics to facilitate widespread adoption and mass production is important for scalability.

- **Cost efficiency**

Although printing technologies are cost-effective, the high cost of advanced materials (e.g., graphene, Ag nanoparticles) can make large-scale deployment expensive.

5.3.2 *Effectiveness Challenges*

Many flexible sensors rely on high-quality conductive inks that are relatively expensive, making up a significant portion of the overall manufacturing cost. Some flexible substrates can also be costly, particularly when used in high-performance or biocompatible applications.

Many printed sensors require additional post-processing steps such as curing, sintering, encapsulation, or lamination to improve performance and reliability. These steps can add to the overall production cost and complexity. Defects or inconsistencies during the printing process can lead to a low yield, increasing costs due to wasted materials and the need for rigorous quality control.

5.3.3 *Strategies for Achieving Scalability and Cost-Effectiveness*

Inks based on low-cost conductive materials can reduce costs. Carbon-based inks are cheaper alternatives to Ag or Au inks while providing good electrical conductivity and flexibility. Organic conductive polymers, like PEDOT:PSS, offer good electrical performance at a lower cost and can be processed using standard printing techniques. Recycling and reusing materials have an impact on the environment. Developing processes to recycle functional inks or recover substrates can help reduce material costs and improve overall cost-effectiveness, with increments upon environmental sustainability.

R2R printing is a continuous, high-speed process that allows for the rapid production of flexible sensors on a roll of flexible substrate. R2R printing significantly improves scalability and reduces costs by enabling high-throughput manufacturing. Gravure and flexographic printing techniques are also capable of high-speed, large-area printing with good resolution and uniformity. They offer high material efficiency, reduced ink wastage, and the ability to print fine features, making them cost-effective for mass production. Inkjet printing is ideal for rapid prototyping, small-batch production, and customization due to its versatility and digital control. It enables low-cost prototyping and small-scale production without the need for expensive masks or molds, making it cost-effective for customized applications. Adopting low-temperature curing techniques, such as photonic curing or UV-assisted sintering, allows the use of cheaper, heat-sensitive substrates and reduces energy consumption, making the overall process more cost-effective. Combining printed electronics with conventional silicon-based components can create a balance between flexibility and functionality, allowing cost-effective integration without compromising performance.

Using automated optical inspection (AOI) systems, in-circuit test (ICT), machine learning, and AI-driven quality control can help identify defects early in the production process, reducing wastage, and improving production yield.

5.4 Environmental Sustainability of PFS

Environmental sustainability is a major concern, as PFS are increasingly being used in various applications. Adopting eco-design approaches, eco-efficient manufacturing, and complete lifecycle analysis is decisive for a reduced environmental impact and footprint and social responsibility. Comparing to rigid sensors, PFS inherently hold strong sustainability potential due to materials savings, energy efficiency, and scalability. However, achieving true sustainability requires deliberate attention to materials selection, design considerations, including lifecycle management and recyclability, and low consumption energy processes, thereby ensuring minimal environmental impacts and maximizing socioeconomic benefits.

5.4.1 Current Sustainability Issues

The environmental impact of current PFS is high, mainly due to the use of non-renewable and hazardous materials (e.g., substrates, inks, solvents), the manufacturing processes are of high energy consumption and waste generation and rise of end-of-life issues. Traditional conductive inks, such as Ag, Au nanoparticle inks, are not only expensive but also rely on non-renewable resources. The extraction and production processes for these metals have significant environmental footprints, involving energy-intensive mining, refining, and processing. Also, commonly used flexible substrates like PET, PEN, and PI are derived from petrochemical sources. Their production is energy-intensive, and they are not biodegradable, contributing to plastic waste.

Printing techniques can result in significant material wastage, particularly when dealing with expensive inks or substrates, leading to inefficiencies and increased environmental impact. Manufacturing processes, such as curing, sintering, or annealing conductive inks, often require high temperatures, leading to substantial energy consumption, especially in large-scale production.

Discarded sensors contribute to the growing problem of electronic waste (e-waste), which is a major environmental concern. Traditional PFS are often challenging to recycle due to the use of mixed materials, making it difficult to separate and reclaim valuable components. Some conductive inks contain toxic metals (e.g., lead, cadmium), which pose environmental and health risks if not properly managed during disposal.

5.4.2 Strategies for Enhancing Sustainability

- **Eco-friendly materials**

The use of biodegradable substrates improves sustainability. For instance, cellulose-based substrates or paper substrates as alternatives to petrol-based polymeric substrates offer a biodegradable option, reducing long-term environmental impact. Polylactic acid, PLA, derived from renewable sources like cornstarch, sugarcane, or bacteria, is another biodegradable substrate that can be used in PFS, promoting a more sustainable approach. The availability of green conductive inks is also a more sustainable approach. Carbon materials offer a more sustainable alternative to metal-based inks. These materials are abundant, non-toxic, and have excellent electrical properties. Organic conductive polymers, such as PEDOT: PSS, are more environmentally friendly than metal-based inks and can be processed at lower temperatures, reducing energy consumption. Shifting from solvent-based to water-based inks reduces the release of volatile organic compounds during printing, leading to a safer, more eco-friendly manufacturing process.

However, many eco-friendly materials may not match the performance, electrical conductivity, or durability of traditional materials, making it challenging to achieve the desired sensor performance while ensuring sustainability.

- **Energy-efficient manufacturing processes**

R2R printing is a highly efficient, continuous process that minimizes material wastage and reduces energy consumption by enabling large-scale production at lower costs. This approach enhances both the scalability and sustainability of PFS manufacturing. The availability of low-temperature processing allows for the use of low-temperature-resistant substrates. Also, techniques like photonic curing, laser sintering, or UV curing allow for low-temperature processing, significantly reducing energy consumption and enabling the use of heat-sensitive, eco-friendly substrates. Direct digital printing techniques (e.g., inkjet and aerosol jet printing) allow precise material deposition, reducing wastage and improving material efficiency, contributing to a more sustainable production process.

- **Recycling and reusability**

Adopting design for disassembly and for recycling principles is a route for enhanced sustainability. Creating sensors with materials that can be easily separated or disassembled at the end of their lifecycle facilitates recycling and reclamation of valuable components, reducing e-waste. Developing inks that can be dissolved, recovered, or reprinted onto new substrates promotes a circular economy, reducing the demand for raw materials and minimizing waste. Developing efficient recycling infrastructure for PFS is still in its early stages, making it difficult to recycle materials at the end of their lifecycle.

- **Integration of renewable energy sources**

Integrating energy-harvesting solutions (e.g., piezoelectric, thermoelectric, or solar energy harvesting) into flexible sensors reduces the need for external power sources (self-powered sensors), improving energy efficiency and sustainability. Combining PFS with printed organic solar cells allows sensors to be powered by renewable energy, making them more sustainable, especially in remote or off-grid applications.

RF SAW sensors require minimal power for operation, as they are often passive (wireless and battery-free) and harvest energy from incoming RF signals. This reduces reliance on batteries and the environmental impact of battery production and disposal.

5.4.3 *Future Trends in Sustainable Printed Flexible Sensors*

Using bioinspired designs and bioderived materials, such as silk fibroin or chitosan, can lead to sensors that are biodegradable, biocompatible, and sustainable, particularly for medical, health, or environmental applications. Adopting circular economy models, where sensors are designed for easy recycling, reuse, or refurbishment, can minimize waste and reduce the need for raw materials, making sensor production more sustainable in the long run. Developing industry-wide sustainability standards and certifications for PFS can encourage manufacturers to adopt eco-friendly practices and promote transparency regarding the environmental impact of their products.

Sustainability is a major consideration for the future of PFS, highly aligned with the growing demand for eco-friendly and resource-efficient technologies. By adopting biodegradable materials, energy-efficient manufacturing processes, recycling initiatives, and renewable energy integration, the industry can minimize its environmental footprint, contributing to a more sustainable future. Although challenges remain in balancing performance, cost, and sustainability, ongoing research and innovation are paving the way for more environmentally responsible PFS that can be widely adopted.

5.5 Integration with AI and IoT

The integration of PFS with AI and IoT is creating smarter, more adaptive, and connected systems that can respond to changing environments, in real time. Smart connected products (Fig. 5.1) are invading our daily life. This convergence is enabling a wide range of applications, by combining the data-gathering capabilities of sensors with the analytical power of AI and the connectivity of IoT networks. The product has embedded sensing (with PFS), local processing (if required, local edge computing AI), and communications (e.g., 5G) capabilities. Products can communicate between them (P-P) or with the cloud-based IoT (P–C), where data analytics may employ AI tools.

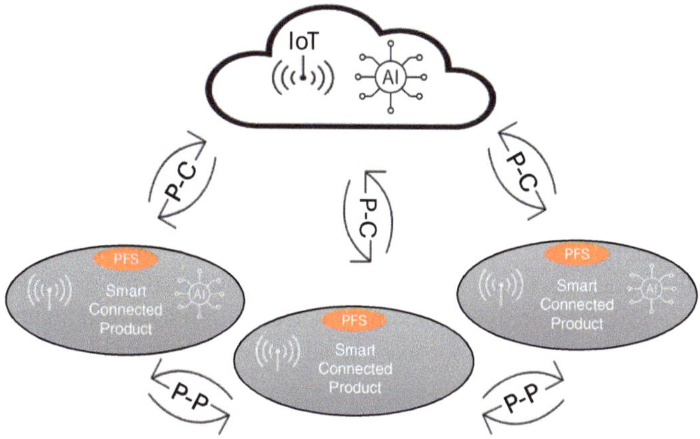

Fig. 5.1 Smart connected products (with embedded PFS, communications and AI) connected to the cloud (IoT and AI): P–P—product-to-product; P–C—Product-to-Cloud

5.5.1 *How PFS, AI, and IoT Work Together*

PFS can work as data sources, as conventional sensors, embedded in edge devices (Fig. 5.2). PFS serves as the foundational layer in this integration, capturing data such as temperature, pressure, humidity, motion, strain, and biochemical markers from the surrounding environment. Their lightweight, bendable, and stretchable nature makes them ideal for embedding in diverse applications, from wearable devices to industrial machinery and smart infrastructure.

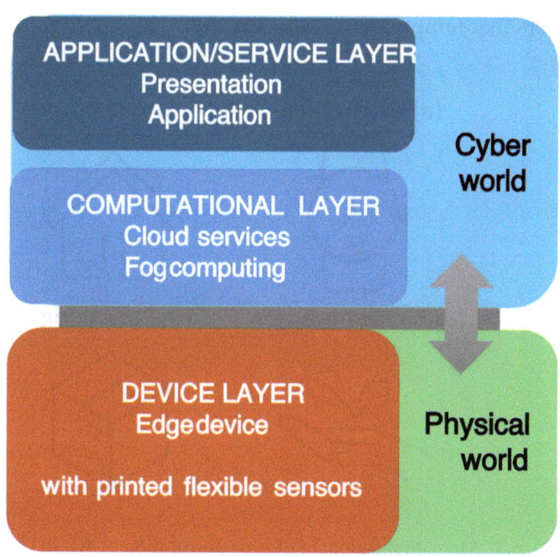

Fig. 5.2 Cyber-physical system with PFS

The IoT provides the connectivity framework that enables flexible printed sensors to transmit real-time data to cloud-based platforms, local servers, or edge devices. IoT protocols such as Bluetooth Low Energy (BLE), Wi-Fi, LoRa, NB-IoT, and 5G facilitate wireless communication. This connectivity allows continuous monitoring, remote data access, and the creation of vast sensor networks, making it possible to track environmental, health, or industrial parameters over wide areas.

AI algorithms process the data collected by PFS, extracting meaningful insights, identifying patterns, and supporting decision-making actions, such as in predictive or autonomous decisions. AI enhances the functionality of sensor systems by providing advanced analytics, anomaly detection, predictive maintenance, and intelligent control, enabling responsive and adaptive systems.

5.5.2 *Key Applications of AI and IoT Integration with PFS*

The integration of PFS with AI and IoT will foster several applications in several fields (Fig. 5.3). PFS embedded in wearables (e.g., smart patches, clothing, wristbands) can monitor physiological parameters like heart rate, blood pressure, body temperature, glucose levels, and sweat composition. The data is transmitted via IoT to a smartphone or cloud platform, where AI analyzes the data to provide personalized health insights. AI algorithms can detect irregular patterns (e.g., arrhythmias, dehydration, glucose spikes) and provide real-time alerts or predictive health recommendations, enabling proactive health management. Some examples are: continuous glucose monitoring for diabetics, smart fitness wearables for athletes, and early detection of cardiac issues.

PFS integrated into building materials or smart home devices can monitor parameters like temperature, humidity, light, air quality, and occupancy. IoT connectivity

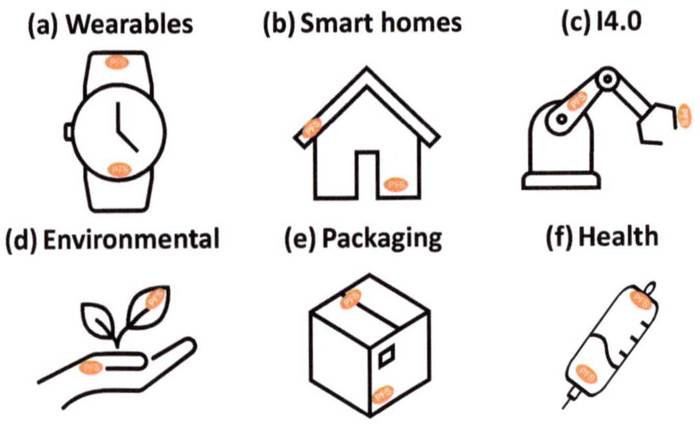

Fig. 5.3 Main applicational fields for integration of AI, IoT, and PFS

enables real-time data transmission to a central control system. AI algorithms can analyze this data and optimize energy usage, adjust HVAC systems, monitor air quality, and detect potential issues (e.g., water leaks, structural damage). Some applications can be envisaged such as smart thermostats that learn occupant preferences, intelligent lighting systems, predictive maintenance for building infrastructure.

In manufacturing and industrial settings, unleashing Industry 4.0 paradigm, PFSs are able of monitoring machinery parameters such as vibration, temperature, pressure, and strain, providing continuous feedback on equipment condition. IoT networks transmit these data to a central platform. AI analyzes sensor data to detect anomalies, predict failures, and optimize maintenance schedules, reducing downtime and maintenance costs (e.g., equipment predictive maintenance, real-time monitoring of conveyor belts). PFS can also monitor processes providing data (such as pressures, temperatures, times, defects) that can be used for their optimization and for improved product quality (e.g., reduced process cycle times, reduced flaws increasing yield, quality control in production lines).

PFS deployed in outdoor environments can monitor parameters such as soil moisture, temperature, humidity, air quality, and pollutant levels. IoT networks relay this data to cloud platforms for analysis. AI models can analyze environmental data to forecast weather patterns, detect pollution hotspots, optimize irrigation schedules, and monitor crop health, enabling precision agriculture and environmental conservation. Some examples can be enumerated: smart irrigation systems, air quality monitoring in cities, and forest fire detection systems.

PFS embedded in packaging can monitor product conditions (e.g., temperature, humidity, impact) throughout the supply chain. IoT connectivity provides real-time tracking and data transmission. AI can analyze this data to optimize logistics, ensure product quality, detect spoilage, and predict delivery times, enhancing supply chain efficiency (e.g., real-time monitoring of perishable goods, pharmaceutical supply chain tracking, intelligent inventory management, and high-value goods protection).

The integration of PFS in healthcare enables real-time monitoring of vital signs (e.g., heart rate, skin properties), medical conditions, and therapeutic environments. When combined with IoT connectivity and AI analytics, these sensors enhance patient care, streamline diagnostics, telemedicine, and improve treatment outcomes. Embedded in wearable devices, patches, or medical packaging, PFS can provide continuous data for personalized healthcare and early intervention.

5.5.3 *Advantages of Integrating AI and IoT with PFS*

As already mentioned, integrating AI and IoT with PFS shows several advantages:

a. **Real-Time Monitoring and Analytics**: The combination of IoT connectivity and AI enables continuous, real-time monitoring of various parameters, allowing for immediate responses to changes or abnormalities.

b. **Predictive Insights**: AI algorithms can identify patterns, trends, and anomalies in sensor data, providing predictive insights that enable proactive decision-making and maintenance.
c. **Scalability**: IoT networks allow for the deployment of large-scale sensor systems that can cover extensive areas, while AI ensures efficient processing and analysis of the vast amount of data generated.
d. **Adaptability and Automation**: AI-driven sensor systems can adapt to changing conditions, learn from historical data, and provide automated responses, reducing the need for human intervention.

5.5.4 Challenges in Integrating AI, IoT, and PFS

Integrating AI, IoT, and PFS offers transformative potential across various domains. PFS may face issues like signal drift, noise, or degradation over time, affecting data accuracy. AI algorithms can compensate for sensor inaccuracies through calibration, noise filtering, and self-learning mechanisms, ensuring reliable data analysis. PFS generates massive amounts of data, which can be challenging to manage, store, and process, especially in large-scale deployments. Edge computing can help process data closer to the source, reducing latency and bandwidth requirements, while AI algorithms can filter and prioritize relevant data. Real-time applications require low-latency data processing, but transmitting sensor data to cloud-based AI systems introduces delays. Deploying AI at the edge for faster processing is possible but it is resource-intensive and requires optimized hardware and algorithms. Reliable and continuous connectivity is critical for IoT-enabled systems. Poor network infrastructure or disruptions can hinder data transmission. Also, large-scale deployments of PFS with IoT can strain network bandwidth, especially for high-resolution or continuous data streams. Ensuring continuous operation of IoT-connected sensors, especially in remote or wearable applications, can be challenging due to power constraints. Incorporating energy-harvesting technologies (e.g., solar, piezoelectric, thermoelectric) and optimizing AI algorithms for low-power processing can enhance energy efficiency. The transmission and storage of sensitive data (e.g., personal data) raise concerns about data security and privacy. Implementing encryption, secure communication protocols, and AI-driven cybersecurity measures can help protect data integrity and confidentiality. The lack of universal standards for IoT devices, communication protocols, and sensor integration complicates seamless interoperability in diverse ecosystems. Different PFS technologies and IoT platforms may not communicate effectively, requiring customized solutions for integration. Addressing these challenges will unlock the full potential of integrating AI, IoT, and PFS for transformative applications across industries.

5.5.5 *Future Trends in AI and IoT Integration with PFS*

The integration of PFS with AI and IoT is revolutionizing how data is captured, analyzed, and used across various industries. Some trends can be identified:

- **Edge AI**: The integration of AI at the edge (e.g., on microcontrollers or edge devices) allows real-time data processing without the need for cloud connectivity, reducing latency and bandwidth usage.
- **Federated Learning**: AI models that learn from decentralized data sources, without sharing raw data, can enhance privacy and efficiency in IoT-connected sensor networks.
- **5G and Beyond**: The widespread of 5G networks will enable faster and more reliable communications for IoT-connected sensors, supporting real-time AI analytics and more sophisticated sensor applications.
- **Energy Harvesting**: Technologies like solar, thermal, or kinetic energy will power PFS and communications, toward autonomous, battery-free, self-powered sensors, extending product lifetime.
- **Digital Twins**: The creation of digital twins (virtual replicas of physical assets) using data from PFS allows for real-time monitoring, simulations in-the-loop, and predictive maintenance in industries such as manufacturing, healthcare, and infrastructures.

This convergence enables real-time monitoring, predictive insights, automation, and adaptive control, driving advancements in innovative applications. As technologies like edge AI, energy harvesting, and 5G continue to evolve, the capabilities and impact of PFS will only expand, ushering in a new era of intelligent, connected systems.

5.6 Emerging Trends and Innovations

PFS technology is rapidly evolving, with numerous emerging techniques and innovative applications. New inks, improved substrates, better and novel printing techniques allow PFS to perform better, being more sensitive and durable. Printed flexible electronics is expanding to print other components besides sensors, such as flexible displays, flexible memories and logic circuits, flexible microprocessors, flexible batteries, and supercapacitors. The integration of PFS with other emerging technologies (e.g., IoT, AI), will provide new breakthrough applications.

- **Materials**

Nanomaterial-based PFS combine the advantages of nanomaterials (high surface area, sensitivity, mechanical flexibility, and unique electronic properties) with scalable printing technologies for high-performance devices. **Carbon allotropes**, such as graphene, graphene oxide (GO), reduced graphene oxide (rGO) and CNT offer excellent flexibility, mechanical durability, sensitivity and high electrical conductivity and low cost, being increasingly used in inks. Other **2D materials** like transition metal dichalcogenides (TMDs) (MoS_2, WS_2, $MoSe_2$, WSe_2) and MXenes are also being investigated. TMDs have semiconducting properties, good mechanical flexibility, and strong photoresponse, being sensitivity to chemical/biological analytes. MXenes show excellent electrical conductivity, hydrophilicity, chemical stability, tunable surface chemistry. **Quantum dots** (QD) are nanoscale semiconductors that exhibit unique optical and electronic properties due to quantum confinement effects. Cadmium-free QD (ZnSe, ZnS, InP, Ag_2S), perovskite QD, and carbon QD significantly enhance sensitivity, specificity, selectivity, tunable optical properties, and multifunctionality, enabling novel sensing applications.

Recent advancements in **transparent conductive oxides** (TCO) significantly enhance the capabilities of printed flexible sensors, combining high transparency and electrical conductivity with superior flexibility. Emerging TCO like Indium Zinc Oxide (IZO), Aluminum-Doped Zinc Oxide (AZO), and Gallium-Doped Zinc Oxide (GZO) are finding applications in wearables, flexible optoelectronic sensors, and environmental and chemical sensors.

The manufacture of PFS with **eco-friendly materials**, with the lowest amounts of materials reduces e-waste. The use of **biodegradable and recycled materials** and of low-energy consumption processes contributes for a more sustainable future.

Self-healing materials can repair themselves after damage (e.g., dents, cracks, or cuts) extending their lifespan and reliability. These materials that can autonomously restore their structure when damaged, often triggered by heat, light, or chemical reactions. These include intrinsically self-healing polymers, self-healing conductive composites, ionic hydrogels, and organohydrogels, which can be used in self-healing inks and self-healing substrates.

- **Printing technologies**

Advanced printing techniques and manufacturing innovations will drive developments on better PFS. **Precision and 3D deposition** of inks in 3D surfaces is possible due to advancements of novel printing techniques such as Aerosol Jet Printing (AJP), which allows printing sensors with high-resolution and complex patterns.

3D printing has rapidly evolved from a prototyping tool to a core technology for fabricating high-performance, flexible, and multifunctional sensors. These technologies enable manufacturing PFS with complex geometries, multimaterial integration, and embedded functionalities. No masks or etching is required, and an efficient use of materials is achieved.

In-mold electronics (IME) is a process where a conformal functional printed polymeric substrate (with printed sensors and electronic circuits) is integrated into a polymeric product during molding process, creating a seamless electronic function (e.g., buttons, displays). This concept can also be applied to 3D printing parts.

Shared manufacturing platforms and technological research infrastructures can help advancing the development of sustainable flexible electronics, providing access to state-of-the-art equipment to smaller companies and to an interdisciplinary approach to printed flexible electronics research projects, fostering innovation in the field.

- **Sensor design and integration with other technologies**

Nature-inspired designs can mimic multifunctional sensing mechanisms in Nature (e.g., of biological processes). This aims at improving sensors performance and functionality. A particular example is e-skin applications, where printed flexible and stretchable sensors and electronics mimic human skin sensing mechanisms. Another example is the development of printed neuromorphic IC capable of mimicking biological neural systems for applications in flexible robotics, e-skin, artificial intelligence, and biomedical devices.

Flexible Hybrid Electronics (FHE) integrates PFS with conventional silicon-based components on flexible substrates for building a full operational flexible device (e.g., with signal processing, data storage, wireless communication, and battery). Developments in bonding rigid components to flexible substrates are required, together with solutions for the thermal and mechanical mismatches between flexible and rigid materials, for more durable products. For example, wireless communication technologies (NFC, RFID) allow flexible sensors to transmit data. Printed antennas are already widely used, but integrating BLE or 5G modules is still a challenge for real-time data transmission.

Energy-harvesting technologies (e.g., triboelectric, piezoelectric, solar) will allow implementing self-powered PFS. These sensors will reduce operational costs by eliminating the use of batteries or enlarging the time for their substitution, reducing the dependence upon external power sources and allowing continuous, remote operations and, eventually, lighter wearable applications.

The **integration of artificial intelligence (AI)** and machine learning (ML) algorithms with printed flexible sensors (PFS) significantly enhances sensor performance by improving accuracy and enabling intelligent calibration through advanced data analysis. AI algorithms can effectively address common challenges such as signal drift, environmental noise, and material-induced nonlinearities, resulting in greater reliability and precision in real-time applications. In addition, machine learning supports predictive maintenance by identifying early indicators of sensor degradation or failure. This enables timely, proactive repair or replacement, reducing system downtime and extending the operational lifespan of the sensors. Looking ahead, the convergence of AI with smart materials used in PFS has the potential to drive the development of next-generation sensors capable of dynamically adapting to changing environments and fine-tuning their sensitivity based on specific applications or contextual demands.

- **Trending Applications**

Flexible and stretchable sensors

Flexible sensor arrays and electronic skins (e-skins) mimic the sensing capabilities of human skin by integrating multiple flexible sensors that can detect pressure, temperature, strain, humidity, and more across a large surface area. E-skins made from nanomaterials, piezoresistive or capacitive elements, and stretchable conductive inks offer high sensitivity, flexibility, and multifunctional sensing capabilities. Stretchable sensors are designed using intrinsically stretchable materials, allowing them to maintain functionality even when stretched, twisted, or bent. Liquid metal inks (e.g., gallium-indium alloy) and conductive polymers are being developed to create highly stretchable and conductive pathways. Future trends involve creating fully stretchable sensors that can be integrated into soft robotics, artificial skin, and other applications requiring extreme flexibility.

Self-healing sensors

Self-healing sensors are capable of autonomously repairing physical or functional damage, extending their lifespan and improving reliability. Strain and pressure sensors used in soft robotics and e-skin can recover sensing functionality after being cut or torn. Temperature sensors may use materials that self-heal while maintaining thermal conductivity. Chemical and biosensors may employ reversible chemical interactions to enable recovery after exposure or contamination. Nevertheless, repeated damage and healing can degrade material properties over time. Ensuring long-term functionality remains difficult. Also, some systems take hours or even days to fully heal, and not all can restore 100% of original performance.

Integration with wearable and IoT devices

The integration of PFS into textiles and e-health monitoring devices will enable continuous, real-time health tracking, enhancing preventive healthcare and fitness applications. The rise of IoT will drive the demand for PFS that can communicate wirelessly, leading to smarter homes, cities, and industries. As IoT expands, the demand for cost-effective, scalable sensors for smart homes, cities, and industrial applications will drive advancements in flexible printed sensor manufacturing.

Printed biosensors and lab-on-skin technologies

Printed biosensors can detect biological markers (e.g., glucose, lactate, pH, and enzymes) using flexible, printed electrodes that interact with biological samples such as sweat, saliva, or interstitial fluid. Integration of nanomaterials (graphene, Au nanoparticles) enhances the sensitivity and selectivity of these biosensors. Lab-on-skin refers to the integration of PFS directly onto the skin, enabling real-time, continuous monitoring of physiological parameters. Combining microfluidics with printed sensors allows the collection and analysis of sweat or other biofluids for monitoring hydration, glucose levels, electrolyte balance, and more.

The future of **printed flexible sensors** is incredibly promising, as they are at the core of next-generation electronics that are lightweight, low-cost, stretchable, and easily integrated into everyday objects. PFS are evolving into ultra-low-cost ubiquitous tools for several applications. In the near future, they will be seamlessly embedded in our clothes, on our skin, in our food packages, and even built into roads and walls, quietly collecting data and making life more connected, safer, and better.

References

1. A. Nag, S. C. Mukhopadhyay, J. Kosel, *Printed Flexible Sensors: Fabrication, Characterization and Implementation*, Springer Cham, 2019.
2. C. Tong, *Advanced Materials for Printed Flexible Electronics*, Springer Cham, 2022.
3. Allied Market Research, *Printed And Flexible Sensors Market Size, Share, Competitive Landscape and Trend Analysis Report, by Printing Technology, By Type, By Application: Global Opportunity Analysis and Industry Forecast, 2022–2031*, Research, Applied Market, UK, 2023.
4. A. Bastola, et al. *Formulation of functional materials for inkjet printing: A pathway towards fully 3D printed electronics*, Materials Today Electronics, 6, p. 100058, 2023.
5. C. Cano-Raya, Z.Z. Denchev, S.F. Cruz, J.C. Viana, *Chemistry of solid metal-based inks and pastes for printed electronics – A review*, Applied Materials Today, 15, pp. 416–430, 2019.
6. Y. Qin, X. Ouyang, Y. Lv, W. Liu, Q. Liu, S. Wang, *A Review of Carbon-Based Conductive Inks and Their Printing Technologies for Integrated Circuits*, Coatings, 13, pp. 1769–, 2023.
7. W. G. Wong, *Additive Manufactured Electronics: Disrupting the Electronics Industry*, Electronic Design, 15-jan-20225. [Online]. Available: https://www.electronicdesign.com/technologies/industrial/article/55261095/electronic-design-additive-manufactured-electronics-disrupting-the-electronics-industry. [access 1 jun 2025].
8. A. H. Espera Jr., J.R. Dizon, Q. Chen, R.C. Advincula, *3D-printing and advanced manufacturing for electronics*, Prog. Addit. Manuf., 4, pp. 245–267, 2019.
9. S. Kokare, F. M. A. Asif, G. Mårtensson, S. Shoaib-ul-Hasan, A. Rashid, M. Roci, N. Salehi, *A comparative life cycle assessment of stretchable and rigid electronics: a case study of cardiac monitoring devices*, Intern. J. Environmental Science and Technology, 19(4), pp. 3087–3102, 2021.
10. N. Zavanelli, K. Kwon, W-H Yeo, *Printed Strain Sensors for Motion Recognition: A Review of Materials, Fabrication Methods, and Machine Learning Algorithms*, IEEE Open Journal of Engineering in Medicine and Biology, 2023.
11. S.S. Dhanabalan, A. Thirumuruga, "Advances in Flexible and Printed Electronics: Materials, fabrication, and applications", IOP Publishing Ltd, 2023.
12. T. Islam, S.C. Mukhopadhyay, B. Georg, "Systems for Printed Flexible Sensors: Design and Implementation", IOP Publishing Ltd, 2022.

J. C. Viana, *Primer on Printed Flexible Sensors*,
https://doi.org/10.1007/978-3-032-06554-4

MIX
Papier aus verantwortungsvollen Quellen
Paper from responsible sources
FSC® C105338
FSC
www.fsc.org

If you have any concerns about our products,
you can contact us on
ProductSafety@springernature.com

In case Publisher is established outside the EU,
the EU authorized representative is:
Springer Nature Customer Service Center GmbH
Europaplatz 3, 69115 Heidelberg, Germany

Printed by Libri Plureos GmbH
in Hamburg, Germany